SAFETY ENGINEERING AND RISK ANALYSIS 2001

presented at

THE 2001 ASME INTERNATIONAL MECHANICAL ENGINEERING CONGRESS AND EXPOSITION
NOVEMBER 11–16, 2001
NEW YORK, NEW YORK

sponsored by

THE SAFETY ENGINEERING AND RISK ANALYSIS DIVISION, ASME

edited by

BEN J. M. ALE
NATIONAL INSTITUTE OF PUBLIC HEALTH AND THE ENVIRONMENT

THE AMERICAN SOCIETY OF MECHANICAL ENGINEERS
Three Park Avenue / New York, N.Y. 10016

ISBN No. 0-7918-3556-1

FOREWORD

Welcome to the eleventh volume of the series on Safety Engineering and Risk Analysis sponsored by the ASME Safety Engineering and Risk Analysis Division (SERAD). This issue contains the papers presented in the six technical SERAD sessions at the 2001 International Mechanical Engineering Congress and Exposition, held in New York during November 11-16, 2001.

SERAD began its call for papers during November 2000 and is pleased to culminate these efforts with this Proceedings. In response to the wishes of the participants in earlier SERAD sessions, the number of papers was reduced in order to allow more time for presentation and discussion. This volume contains the sixteen papers selected for the 2001 SERAD sessions. The program was developed in support of the Congress highlight topic of "Progress Through Partnerships" and collaborated with the Engineering and Technology Management (E&TM) Group to develop an E&TM symposium on "Partnering to Succeed: Keys to Managing Development, Risk and Globalization."

For SERAD, safety, risk, and reliability continue to be key public issues as the stream of reports about accidents in facilities and in transport continues and national and international projects to improve safety and reliability increase in effort. Our six sessions are:

- The New Frontier of Product Safety at the End of Product Life
- The Global Application of Risk-Informed or Performance-Based Standards
- Discussion Session: Impacts of Standards and Design on Safety
- Beneficial Approaches and Examples of Procedures for Safer Use
- Business Approaches to Managing Risk and Reliability
- Conflict Resolution Between Siting of Facilities and the Public

SERAD thanks its Session Chairs and Co-Chairs who worked hard with the authors in these sessions and who assisted with the coordination of the Peer Reviews of the manuscripts: Ms. Alexis Flippen and Messrs. Dick van den Brand, Gerald Laheij, Robert Navarro, David Pyatt, and Paris Stavrianidis. Our special thanks are extended to William Doerr, the program coordinator of last year's sessions, who was a great support among others with the communications between organizations located at different continents and thus different time zones.

Finally, our thanks are extended to Kenneth Horne, the E&TM Program Coordinator, Hirosh Honda, the E&TM Proceedings Editor, and Jeffrey Rode, the E&TM Group Vice President for their interest.

Ben. J. M. Ale
Proceedings Editor
2001 SERAD Program

CONTENTS

Proceedings of
2001 ASME International Mechanical Engineering Congress and Exposition
November 11–16, 2001, New York, NY
SERA-Vol. 11

IMECE2001/SERA-24000

ON THE SAFETY OF INFEEDING VERTICAL GARDEN SHREDDERS

Dennis B. Brickman
Triodyne Inc.
5950 W. Touhy Ave., Niles, Illinois 60714
(847) 677-4730 Fax (847) 647-2047
(e-mail) brickman@triodyne.com

Ralph L. Barnett
Professor, Mechanical and Aerospace Engineering
Illinois Institute of Technology
Chicago, Illinois 60616

ABSTRACT

This paper addresses consumer garden shredders of the hammermill type with vertical hoppers that are manually fed. The purpose of this paper is to show that garden materials presented to a vertical garden shredder through the inlet hopper will not pull an erectly standing operator's hand into the flails. In order for an erectly standing operator to contact the flails, it is necessary for the shoulder to move downward. Experiments demonstrate that the operator's shoulder moves insignificantly downward during pull-in excursions using various garden materials that are attached to the operator's hand through snagging mechanisms, hand friction, and entanglement.

INTRODUCTION

The principle hypothesis of this paper is that an erectly standing operator's shoulder and torso will not move downward during a pull-in excursion of garden materials into the shredder hopper. An operator's arm is not long enough to touch the flails while maintaining an erect standing position. Dealing first with this aspect of the problem, Fig. 1 defines the geometry of a consumer hammermill chipper/shredder. The cutaway indicates the flails reach their apex at a distance of 63.0 cm (24.8 in.) from the ground. Adults in an erect standing position with dangling arms exhibit a knuckle clearance with respect to the ground. For males and females in the 1, 50, and 99 percentile classes, the ground clearance is shown in Table 1 [1 - 3]. In each case, hand-flail contact will not occur because the clearance is greater than 63.0 cm (24.8 in.), the top of the flails. This clearance is conservative because the serpentine route through the hopper to the flails lengthens the path.

The safety from this geometry will be maintained throughout a pull-in scenario if it can be shown that the shoulder remains stationary or moves upward. Shoulder immobility was in fact demonstrated in seventeen test configurations which varied the initial position of the hand relative to the top of the hopper. Furthermore, various snagging profiles were studied that involved nailing tree branches to leather gloves and enveloping the wrist of an anthropomorphic dummy with garden jute

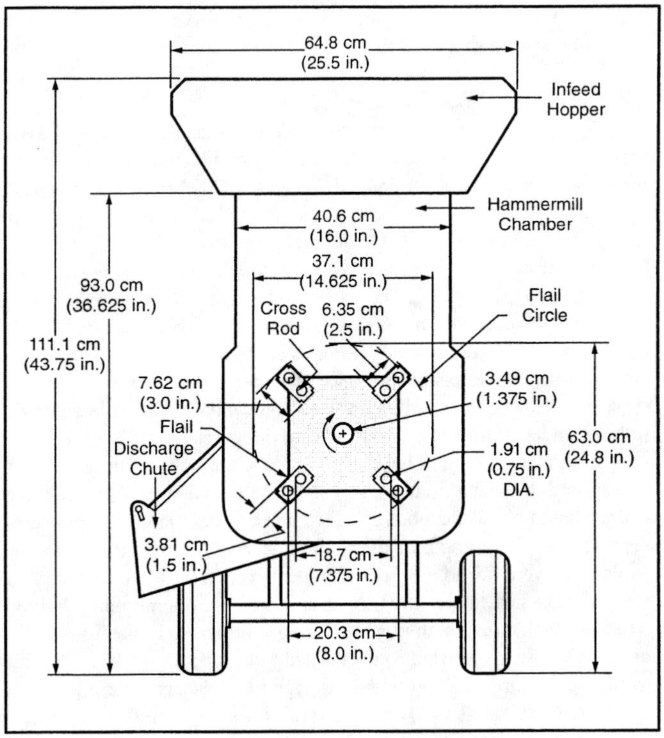

Fig. 1 - Chipper/Shredder Geometry

twine. In every case, the downward shoulder movement was insignificant. This result was also established analytically in reference 4 by modeling the arm as a ballistic pendulum under an impulse loading caused by pulling garden materials through an operator's grip at high speeds [4]. Material is snatched from an operator's hand at the peripheral speed

Table 1 - Adult Knukle Height (1 in. added for shoes)

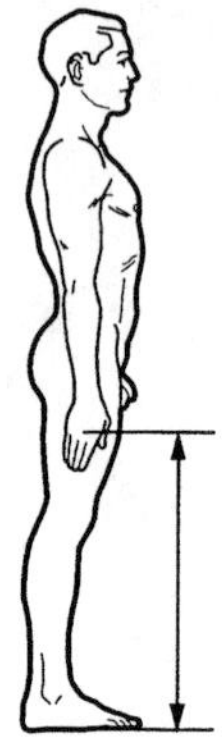

Knuckle Height (in.)

1	Percentile - Male ------ 27.7
50	Percentile - Male ------ 31.0
99	Percentile - Male ------ 34.5
1	Percentile - Female --- 25.9
50	Percentile - Female --- 27.9
99	Percentile - Female --- 32.5

of the hammermill flails, 252.7 km/h (157 mph). When material trails the leading edge of the fist by 30.5 cm (1 ft), the impulse time is approximately 0.00436 sec. When this time is compared to the arm's natural period of vibration, 1.13 sec., the impulse time is approximately 0.386% of the period. Under these circumstances, there is no hand movement until after the vegetation has emerged from the grip. Snagging scenarios provide even smaller impulse times than the friction grip. The impulse analysis indicates that the shoulder actually moves upward during the impulse phase as shown in the appendix of reference 4.

TESTING PROTOCOL

Figure 2 shows the dynamic test setup that was used to study the pull-in phenomenon in a consumer rotating hammermill chipper/shredder. The particular chipper/shredder studied operated at 3600 rpm, weighed 100.7 kg (222 lb), and had its flails removed to preclude premature cutting of the fibers. A 175.3 cm (69 in.) tall anthropomorphic dummy weighing 77.6 kg (171 lb) is posed next to the infeed hopper with its right hand positioned in the central hopper region. The dummy has its feet resting on the ground and its waist supported by an overhead strap which allows the torso to rotate forward if a downward tension is applied to the arm. The support system offers no restoring resistance. Further, in some cases, the pull-in scenario begins with an arm that is bent at the elbow. Others begin with a fully extended arm facing in a downward direction.

In the snagging and hand friction tests, the pull through force was transmitted through a clothesline which was threaded through the infeed hopper, hammermill chamber, and discharge chute of the test chipper/shredder. The tensile strength of the clothesline was 754 pounds and various appliances were tied to its upper end during the tests. In the entanglement tests, garden jute twine was tied around the anthropomorphic dummy's wrist and then threaded through the chipper/shredder. Typically, the chipper/shredder was started in its unclutched state and its engine was brought to full speed; the clutch was then engaged and the hammermill was allowed to reach its full 3600 rpm. The pull-in force was initiated by pulling on the trailing edge of the clothesline or garden jute twine which emerged from the discharge chute. This caused the clothesline to wrap around the crossbars of the hammermill to achieve a pull-in speed of 161.9 km/h (100.6 mph); this speed is 64% of the peripheral speed of the flails. Still photography and videotaping captured the test trials.

Fig. 2 - Dynamic Test Setup

TESTING - HAND GRIP

Test Setup

When vegetation is grasped by a hand, there is a trailing length behind the leading or downward edge of the hand. As the trailing length is pulled through the hand, the resisting friction force causes an impulse to be delivered to the hand. This impulse is equal to the friction force times the pull through time; here, the friction force is equal to two times the coefficient of friction between the vegetation and the gloved or bare hand. A constant force grip of 45 pounds was simulated by using a 45 pound constant force spring illustrated in Fig. 3 with a length of 30.5 cm (1 ft) to represent a vegetation trailing length of 30.5 cm (1 ft). It should be noted that a grip of 45 pounds is much greater than the force required to feed materials into the infeed hopper.

Testing

Four tests were conducted with an erectly standing anthropomorphic dummy using a bent elbow which positioned the hand over the center of the infeed hopper. In each case, the capacity of a constant

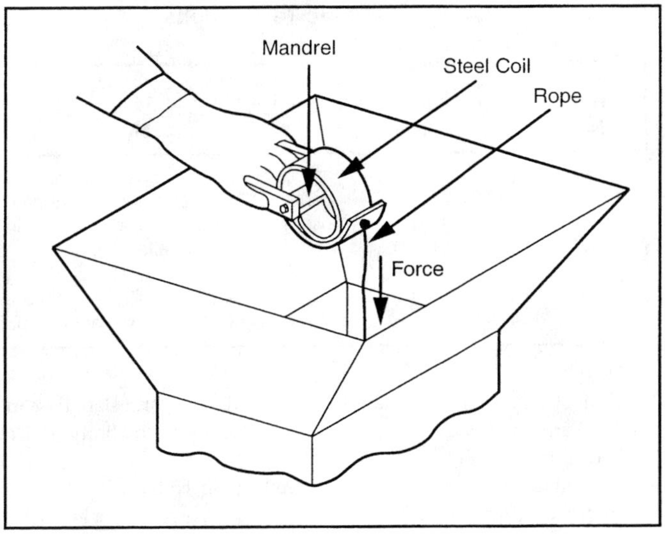

Fig. 3 - Constant Force Spring

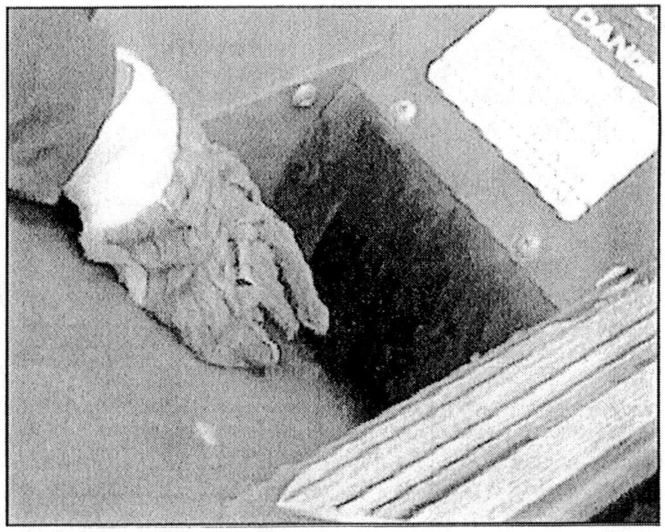

Fig. 5 - Final Hand Position - Constant Force Spring

force spring was verified within very close tolerances to be 45 pounds before attaching it to the dummy's hand. The pull force was initiated and the response of the dummy was recorded. Fig. 4 is a typical critical sequence for one of the tests which shows the unloaded arm in frame 4A, a following frame 4B, where the constant force spring has been pulled through, and frame 4C, which illustrates the final position of the hand, arm, and shoulder after all motion stopped. Figure 5 is a photograph of the final hand position. It can be observed in this sequence that the downward vertical motion of the shoulder is imperceptible. In addition, Fig. 5 indicates that in the final position the hand has not entered the hammermill chamber. The results did not vary in a significant way among the four tests. In all cases, the pull through phenomenon is so fast that the 30.5 cm (1 ft) trailing length of the constant force spring completely passes through the hand in one video frame (1/30 of a second).

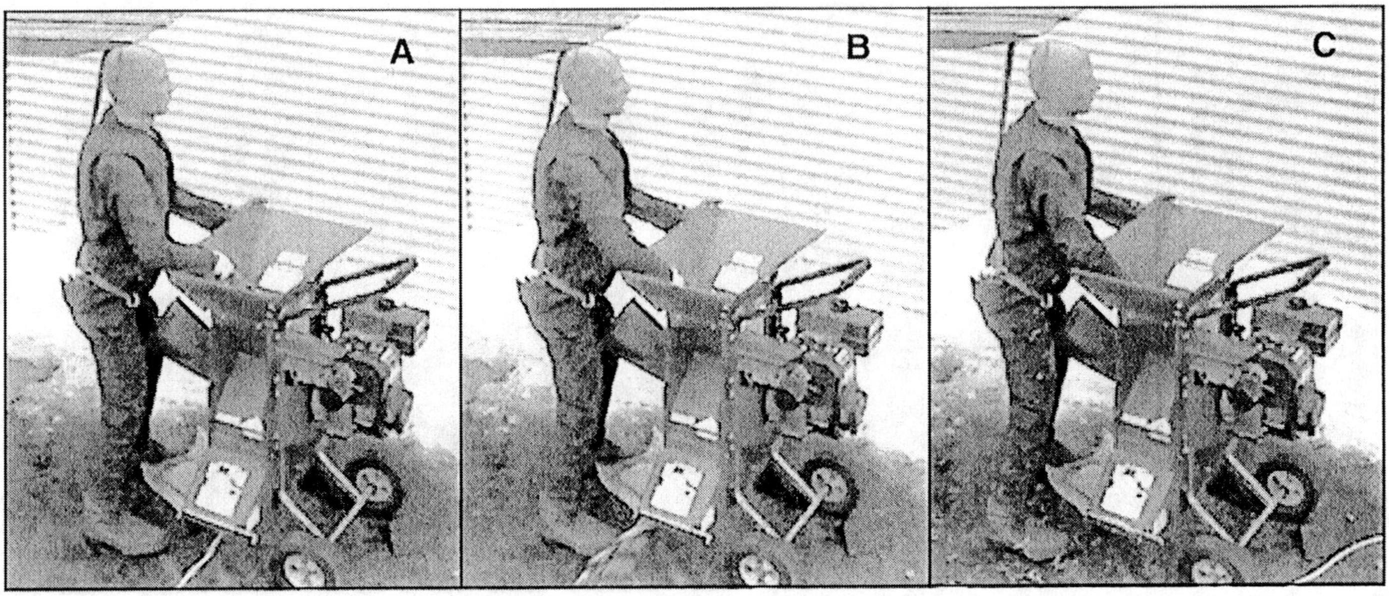

A) Unloaded Arm **B) Constant Force Spring Pulled Through Hand** **C) Final Arm Position**

Fig. 4 - Video Sequence - Constant Spring Force

TESTING - SNAGGED GLOVE

Test Setup

To study the effect of snagging during a feeding scenario, a high strength connection was constructed by nailing a leather palmed glove to a branch with a single or double nail. The branch in turn was fastened to a clothesline which was threaded through the chipper/shredder. Four different species of branches were used . Each test branch was nominally 1.27 cm (0.5 in.) in diameter.

Static Connection Tests

Single or double nailed joints were created by nailing the palm of a new leather glove to branches of four species. Static tests were conducted to determine the tenacity of the connection. In a typical static test, the gloved hand was held in a vise and the nailed branch was pulled horizontally by a rope attached to a winch. Inserted in this pulling line was a load cell connected to a Chatillon gauge which recorded the maximum pull force. Table 2 tabulates the static test results where the failure modes are indicated. The failure modes involved the nails pulling out of the palm of the leather glove or the branches pulling out of the nails; no nail failures were observed. The highest connection resistance was recorded at 122 pounds.

Dynamic Testing

Using the test protocol, dynamic pull through tests were conducted on all four species' branches using single and double nailed joints into 1.27 cm (0.5 in.) diameter branches. In the first seven tests, the anthropomorphic dummy's elbow was bent and the gloved hand was located in the center of the top horizontal plane of the infeed hopper. Three additional tests were conducted with an extended arm dangling in a vertical attitude which brought the hand initially to the bottom of the infeed hopper.

In all ten tests, the joint was severed within 1/30 of a second (one video frame) after initiation of the pull. A typical test using a bent elbow produced the three frame sequence shown in Fig. 6. The first frame, 6A, defines the geometry just prior to loading; the second frame, 6B, represents the following video frame after the connection has been severed; frame 6C depicts the resting position after all motion has ceased. Figure 7 is a photograph of the final hand position. In all cases the downward shoulder movement was insignificant. Also, the bent arm was straightened with the hand remaining outside of the hammermill chamber. In addition to the two modes of failure observed in the static tests, the dynamic tests also included a case where the connection at the glove remained intact and the branch severed.

Table 2 - Static Test Results

Trial No.	Branch No.	Number of Nails	Connection Resistance (pounds)	Failure Mode
1	1	1	78.7	Branch pulled out of nail
2	1	1	86.0	Branch pulled out of nail
3	1	1	42.8	Branch pulled out of nail
4	1	2	122.0	Nails pulled out of glove
5	2	2	40.0	Branch pulled out of nails
6	3	2	32.0	Branch pulled out of nails
7	4	2	35.0	Branch pulled out of nails

A) Unloaded Arm B) Severed Connection C) Final Arm Position

Fig. 6 - Video Sequence - Snagged Glove with Bent Elbow

Fig. 7 - Final Hand Position - Snagged Glove with Bent Elbow

A typical pull through sequence associated with one of the three extended arm tests is depicted in Fig. 8. Once again, frame 8A illustrates the geometry immediately before loading; frame 8B is the video frame following the onset of loading; the final position after all motion has stopped is illustrated in frame 8C. Other than a small change in the hand configuration, the geometry of the anthropomorphic dummy was unaffected by the pull-in scenario. The hand remained outside the hammermill chamber in all three extended arm tests.

TESTING - TIED WRIST

Test Setup

Using garden jute twine tied around the wrist of an anthropomorphic dummy, three dynamic tests were conducted with an extended arm oriented vertically downward into the bottom of the infeed hopper. The position of the dummy followed the test protocol with one exception; twine was used instead of clothesline. Preceding the pull-in tests, a universal testing machine was used to determine the tenacity of 3 samples of the garden jute twine.

Test Results

Static testing provided three garden jute twine sample breaking strengths: 48 pounds, 62 pounds, and 59 pounds. Following these tests, the twine was wrapped and tied around a gloved hand as depicted in Fig. 9. The results of one of the three dynamic pull-in tests is characterized in Fig. 10. Frame 10A defines the geometry of the anthropomorphic dummy before loading occurred; the following frame, 10B, is shown where the connection has been broken; the final geometry after cessation of motion is illustrated in frame 10C. In two of the three tests, the twine failed at the knot; the twine fractured in the third test at an interior location. The resting position of the anthropomorphic dummy's hand in the test sequence is illustrated in Fig. 9 which is typical of all three dynamic twine tests. It should be noted that the hand never entered the hammermill chamber. This is consistent with the observation that the downward shoulder movement was imperceptible during the pull through scenario.

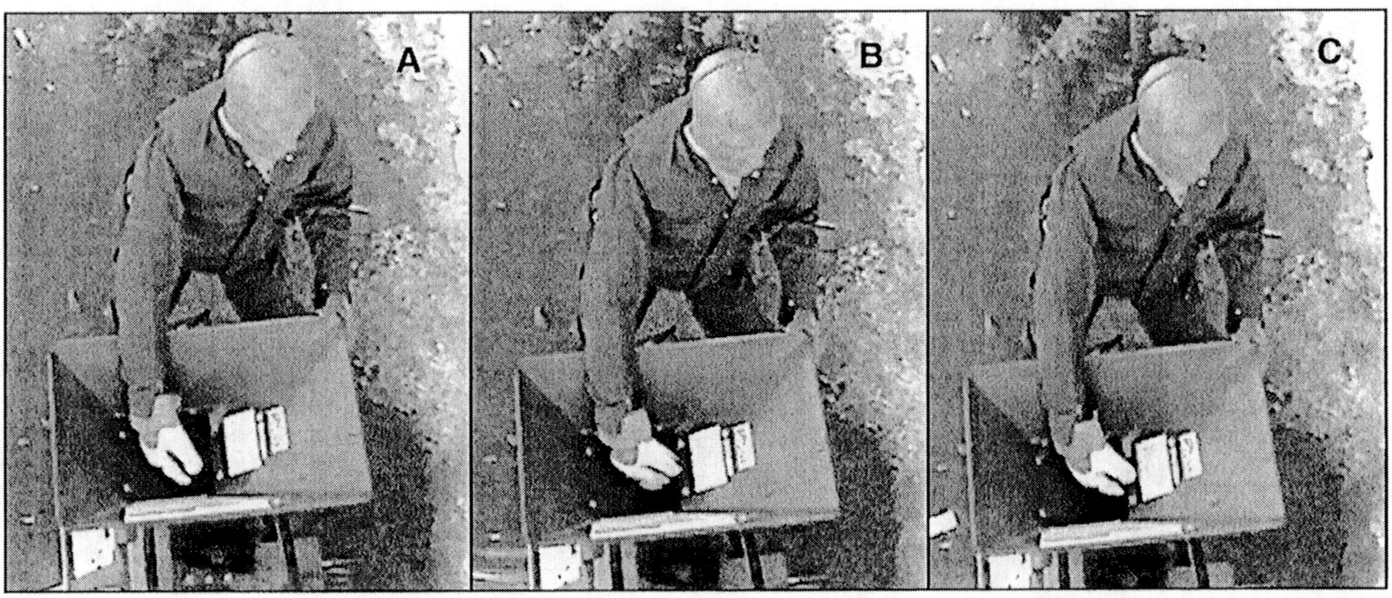

A) Unloaded Arm **B) Severed Connection** **C) Final Arm Position**

Figure 8 - Video Sequence - Snagged Glove with Extended Arm

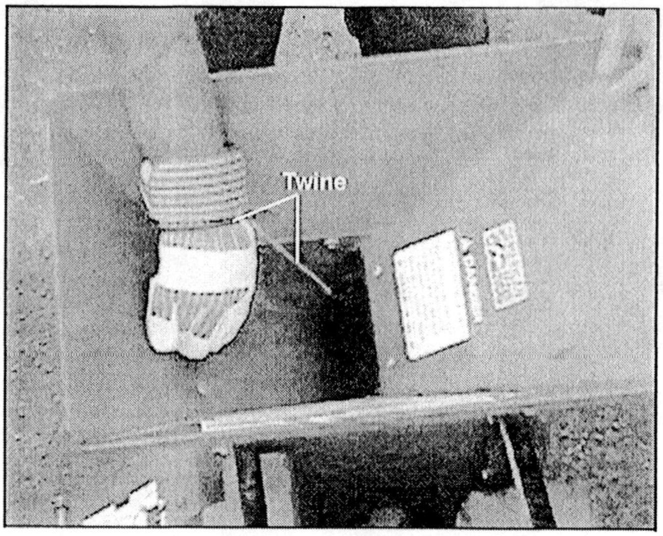

Fig. 9 - Final Hand Position - Tied Wrist

CONCLUSIONS

1. If the shoulder does not move downward appreciably, an erectly standing operator's hand will not make contact with the flails.

2. For impulse testing performed under scenarios of snagging, entanglement, and friction grip, downward shoulder movement was insignificant.

3. The tensile integrity of the vegetation and its connection to the operator's hand must be sufficiently great that it will overcome the inertia of the upper torso before hand-flail contact occurs.

4. It is incumbent upon a technical investigator who formulates a chipper/shredder pull-in hypothesis to prove the hypothesis using testing, dynamic analysis, or computer simulation.

5. Manufacturers of consumer chipper/shredders have admonished users not to reach beyond certain benchmarks in the infeed hopper throat. Warnings and instructions of this type will preclude operators from reaching directly into the flails and represent a sensible warning for controlling the hand-flail contact hazard.

REFERENCES

[1] Damon, A. et al., *The Human Body in Equipment Design.* Harvard University Press, Cambridge, MA 1966, pp. 72 - 73.

[2] Tilley, A. R., *The Measure of Man and Woman: Human Factors in Design.* The Whitney Library of Design, New York, 1993, pp. 11 & 13.

[3] *Anthropometry of Infants, Children, and Youths to Age 18 for Product Safety Design.* SP-450, SAE International, 1977, pp. 292 - 293.

[4] Barnett, R.L., and Brickman, D.B., "Chipper/Shredder: The Pull-in Hypothesis," *DETC 99/RSAFP-8868*, Proceedings of the 1999 ASME Design Engineering Technical Conferences, Las Vegas, NV, Sept. 12 - 15, 1999, pp. 1- 9.

A) Unloaded Arm **B) Broken Connection** **C) Final Arm Position**

Figure 10 - Video Sequence - Tied Wrist

Proceedings of
2001 ASME International Mechanical Engineering Congress and Exposition
November 11–16, 2001, New York, NY
SERA-Vol. 11

IMECE2001/SERA-24001

LADDER RUNG VS. SIDERAIL HAND GRIP STRATEGIES

Ralph L. Barnett
Professor, Mechanical and Aerospace Engineering
Illinois Institute of Technology
Chicago, Illinois 60616

Peter J. Poczynok
Triodyne Inc.
5950 W. Touhy Ave., Niles, Illinois 60714
(847) 677-4730 Fax (847) 647-2047
(e-mail) infoserv@triodyne.com

ABSTRACT

When climbers lose their foothold on fixed, straight or extension ladders, the incipient fall may be arrested by gripping either the ladder rungs or siderails. Grasping the rungs provides an interference or power grip; squeezing the siderails provides a friction grip which is the primary focus of this paper. The falling scenario begins with free fall that lasts for the duration of the simple reaction time. Free fall is then decelerated by contravening friction forces derived from hand grip forces rapidly applied to the siderails. Using hand grip/time histories for various individuals, their fall distances were calculated for bare and gloved hands on a vertical steel fixed ladder. Sometimes the candidates could not arrest their falls; often their fall distance was too great to prevent ground impact. Under some circumstances, the vertical motion was brought under timely control. Although a rich literature is available for characterizing grip strength, data reflecting grip/time profiles does not appear. Grip strength/time diagrams were measured for fourteen test subjects.

INTRODUCTION

When climbing a fixed, extension or straight ladder, the climber may elect to grasp either the siderails or the rungs. A review of the literature relative to "How to Climb a Ladder" reveals that one of the articles recommends holding onto the siderails [1], two advocate holding onto the rungs [2, 3] and the eleven remaining papers admonish the climbers to hold onto the ladder without specifying a handhold preference [4-14]. None of the referenced articles attempt to technically establish which strategy, siderail support or rung support, is better. Such a comparison is the primary goal of this paper.

Under ordinary ladder climbing conditions, users are usually supported by three or four appendages. To prevent falling, the hands must supply a small horizontal resistance to prevent the user from rotating backward off the rungs. It is not necessary for the hands to grip either the rungs or the siderails; the fingers need only provide a "hook". When normal conditions prevail, the two climbing protocols provide equivalent safety. It is only when there is a complete loss of foothold that radical differences between the protocols may be distinguished. The siderail strategy cannot reestablish equilibrium without securing a tight grip on the siderails; the rung strategy need only maintain the hook-like geometry of the fingers to arrest the fall.

Two central concepts are used in this paper to analyze the "siderail" climbing strategy. First, classical equations of motion are formulated to provide fall height, *H*; i.e., the vertical fall distance from the instant foothold is lost until the motion is arrested. Second, a "best case scenario" is always assumed for the siderail climbing protocol. For example, the dominant hand grip strength will be taken for both hands. Assumptions of this kind always lead to predictions for *H* that are less than the actual fall heights.

Hand Grip/Time Relationship

The literature abounds with grip strength studies that reflect the following factors: gender, age and handedness, time of day, body position, altitude, gloves, arm support, grip size, oxygen, temperature, fatigue, diet, training, height and weight, wrist and forearm position and smoking. References to the various hand grip studies may be found in a literature review conducted by Dennis B. Brickman [15]. None of the studies consider the time history of the grip force application. The grip/time history directly effects the retarding force acting on a falling climber who is grasping the siderails.

Grip/time relationships were measured for fourteen candidates using the protocol outlined in Appendix A. Figure 1 is a typical grip/time diagram which is bounded by a bilinear curve; the initial portion is called the onset curve and the horizontal portion is the maximum grip strength curve. The intersection of the two curves defines the onset time for achieving the maximum grip strength. The fictitious bilinear curve underestimates the actual time for achieving maximum grip strength; it provides a constant maximum grip strength throughout a longer period than actually experienced in reality. Thus, the bilinear bounding curve overestimates a climber s ability to decelerate his or her body during a fall.

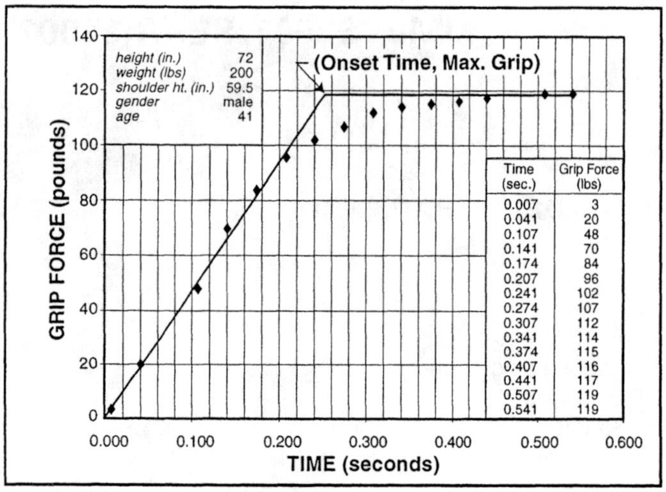

Fig. 1 - Grip/Time Diagram -Test Subject 1

The onset time and the maximum grip strength are tabulated in Table I for 14 test subjects. The complete data base may be found in [16].

Friction

When a climber uses the siderails to arrest a fall on a vertical fixed ladder, each hand produces a gripping force G on two surfaces. If the coefficient of sliding friction between the siderail and the hand is represented by μ_s, the total drag resistance D resisting the fall is:

$$D = 4G\mu_s \qquad \text{Eq. 1}$$

Using the test setup depicted in Fig. 2, the room temperature sliding friction was measured for each of the fourteen test subjects in the grip strength testing program. Candidates rested their hands on a table with their palms facing upward; a Chatillon digital force gauge was fastened to a mild steel angle and the 5.71 pound assembly was dragged over their palms with a standard "slip meter" power winch. A tripod mounted camcorder recorded the drag resistance at each of seven stations one inch apart. The associated average coefficients of sliding friction are tabulated in Table I for the fourteen test candidates. The coefficient of sliding friction is defined as the Drag Force divided by the Normal Force acting on the hands, i.e., μ_s = Drag Force ÷ (5.71 lbs). Typical data tabulations and calculations are shown in Table B-1 for Test Subject 1. The test setup in Fig. 2 was used to determine the average coefficient of friction between a mild steel angle and nine types of work gloves. The results are tabulated in Table II. The complete data base may be found in [16].

The following equipment was used in the test program:

a) Carbon Steel Angle

Cross section:	1-1/2 x 1-1/2 x 1/8 inches
Length:	42-1/4 inches

b) Power Winch

Brand:	Whiteley Industries
Model:	HPS-3
Pull Speed:	3.5 inches/minute

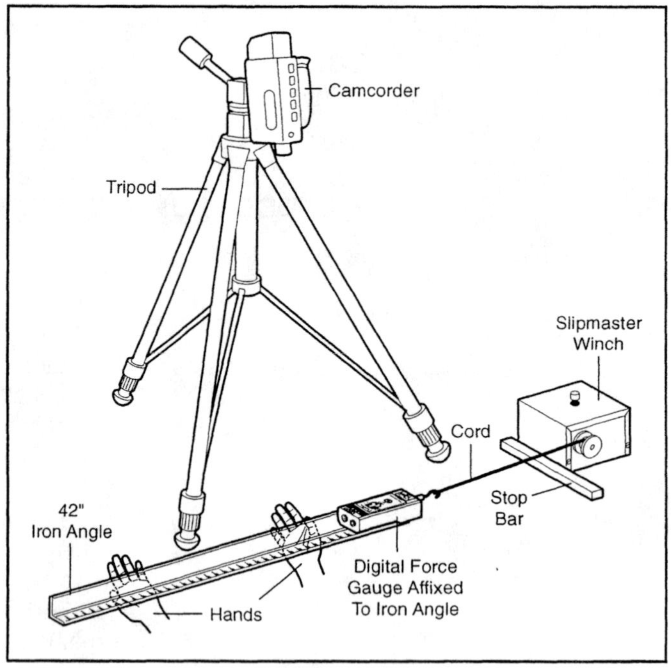

Fig. 2 - Sliding Friction Test Set-Up

c) Chatillon Digital Force Gauge

Model:	DRC 100
Serial:	M002027
Capacity:	100 lbs
Increments:	0.1 lbs

Fall Height - Siderail Strategy

Under the influence of gravity, a climber's loss of foothold results in a falling excursion that can be characterized using Newton's Second Law of Motion, i.e., a particle's acceleration is proportional to the impressed forces and inversely proportional to its mass. Using this law, there are three natural ranges where the downward motion may be studied; free fall, increasing grip resistance and maximum constant grip strength. At the beginning of each range, the clock will be set at zero and the symbol t will represent the time in that range. The overall time will be represented by the term T.

During the initial phase of a fall, sensory receptors are stimulated and after a time interval called the simple reaction time, t_r, [17] the climber's hands will begin to counteract the fall by tightening their grip on the siderails. Referring to the free-body diagram shown in Fig. 3, the equation of motion and the boundary conditions become respectively,

$$\ddot{x} = g \quad 0 \le T \le t_r \qquad \text{Eq. 2}$$

$$At:\ t = 0,\ \ x = \dot{x} = 0 \qquad \text{Eq. 3}$$

where g is the acceleration due to gravity and $\dot{x} \equiv \dfrac{dx}{dt}$.

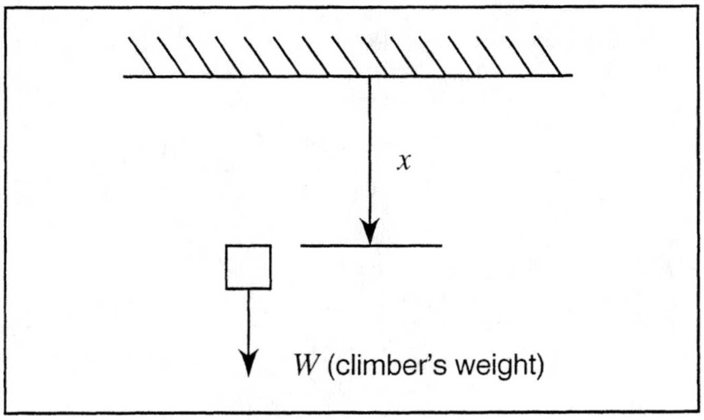

Fig. 3 - Free Body Diagram: $0 \le T \le t_r$

When these equations are solved we obtain the fall height H_1 and the velocity $\dot{x}_1$ at the time $t = t_r$; thus,

$$H_1 \equiv x(t_r) = \frac{1}{2} g t_r^2 \qquad \text{Eq. 4}$$

$$\dot{x}_1 \equiv \dot{x}(t_r) = g t_r \qquad \text{Eq. 5}$$

After a time t_r has elapsed, the hands begin to grasp the siderails with a linearly increasing intensity corresponding to the onset curve represented in Fig. 1. This linear portion of the grip/time diagram terminates at the onset time, t_0, which is shown as the abscissa of the intersection of the two bounding curves. It is convenient to study the second range of motion in the time interval $t_r \le T \le t_0$. A free-body diagram reflecting the prevailing conditions in the second range is depicted in Fig. 4a where the drag force D given by Eq. 1 has the time history shown in Fig. 4b. The maximum grip strength is represented by the term G_{max} and the climber's weight by W. Thus, the equation of motion and the boundary conditions at the beginning of phase two are:

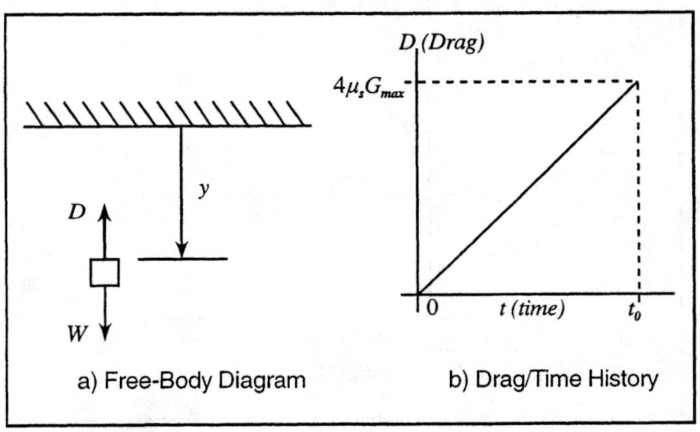

Fig. 4 - Free Body Diagram: $t_r \le T \le t_0$

$$\ddot{y} = -g\left[\frac{4\mu_s G_{max}}{W}\left(\frac{t}{t_0}\right) - 1\right] \qquad \text{Eq. 6}$$

$$At:\ t = 0,\ y = 0 \qquad \text{Eq. 7a}$$

$$t = 0,\ \dot{y} = \dot{x}(t_r) = g t_r \qquad \text{Eq. 7b}$$

Solving these equations we obtain the fall distance H_2 in the second range and the velocity $\dot{y}_2$ at the time t_0 at the end of the second range:

$$H_2 \equiv y(t_0) = \frac{1}{2} g t_0^2 - \frac{2\mu_s g G_{max}}{3W} t_0^2 + g t_r t_0 \qquad \text{Eq. 8}$$

$$\dot{y}_2 \equiv \dot{y}(t_0) = g\left[t_r + t_0\left(1 - \frac{2\mu_s G_{max}}{W}\right)\right] \qquad \text{Eq. 9}$$

In the third and final range of motion, the maximum grip strength of the climber has been achieved which produces a constant drag resistance of $4\mu_s G_{max}$. This case is represented by the free-body diagram shown in Fig. 5. The equation of motion and the appropriate boundary conditions at the beginning of the range are:

$$\ddot{z} = -g\left[\frac{4\mu_s G_{max}}{W} - 1\right] \qquad \text{Eq. 10}$$

$$At:\ t = 0,\ z = 0 \qquad \text{Eq. 11a}$$

$$t = 0,\ \dot{z} = \dot{y}(t_0) = g\left[t_r + t_0\left(1 - \frac{2\mu_s G_{max}}{W}\right)\right] \qquad \text{Eq. 11b}$$

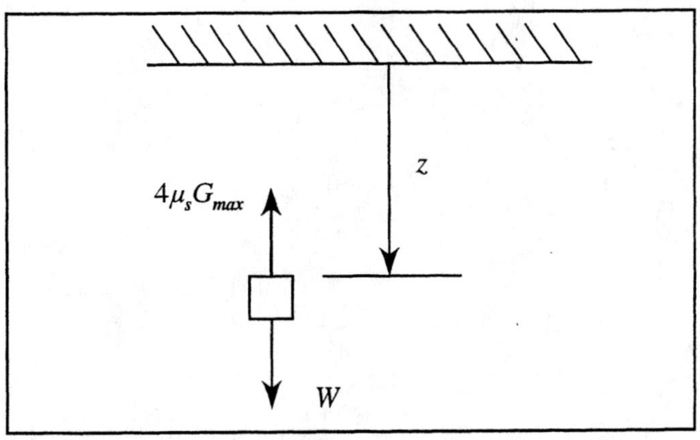

Fig. 5 - Free Body Diagram: $t_r + t_0 \le T$

These equations define the following solutions:

$$z = \frac{1}{2} g t^2\left(1 - \frac{4\mu_s G_{max}}{W}\right) + g t\left[t_r + t_0\left(1 - \frac{2\mu_s G_{max}}{W}\right)\right] \qquad \text{Eq.12}$$

$$\dot{z} = g t\left(1 - \frac{4\mu_s G_{max}}{W}\right) + g\left[t_r + t_0\left(1 - \frac{2\mu_s G_{max}}{W}\right)\right] \qquad \text{Eq.13}$$

The maximum fall in the third range, H_3, occurs when the climber's velocity $\dot{z}$ becomes zero. The associated time, t_m, is found by setting Eq. 13 equal to zero; hence,

$$t_m = \frac{\left[t_r + t_0\left(1 - \frac{2\mu_s G_{\max}}{W}\right)\right]}{\left(\frac{4\mu_s G_{\max}}{W} - 1\right)} \qquad \text{Eq. 14}$$

Substituting Eq. 14 for t in Eq. 12 gives,

$$H_3 = z(t_m) = \frac{1}{2} g \frac{\left[t_r + t_0\left(1 - \frac{2\mu_s G_{\max}}{W}\right)\right]^2}{\left(\frac{4\mu_s G_{\max}}{W} - 1\right)} \qquad \text{Eq. 15}$$

The total fall height of the climber, H, is the sum of the fall heights in the three ranges; thus,

$$H = H_1 + H_2 + H_3 \qquad \text{Eq. 16}$$

$$H = \frac{1}{2} g t_r^2 + \left[\frac{1}{2} g t_0^2 \left(1 - \frac{4\mu_s G_{\max}}{3W}\right) + g t_r t_0\right] + \frac{1}{2} g \frac{\left[t_r + t_0\left(1 - \frac{2\mu_s G_{\max}}{W}\right)\right]^2}{\left(\frac{4\mu_s G_{\max}}{W} - 1\right)} \qquad \text{Eq.17}$$

or

$$H = \frac{1}{2} g t_r^2 \left\{1 + \left(\frac{t_0}{t_r}\right)^2 \left(1 - \frac{4\mu_s G_{\max}}{3W}\right) + 2\left(\frac{t_0}{t_r}\right) + \frac{\left[1 + \left(\frac{t_0}{t_r}\right)\left(1 - \frac{2\mu_s G_{\max}}{W}\right)\right]^2}{\left(\frac{4\mu_s G_{\max}}{W} - 1\right)}\right\} \qquad \text{Eq. 18}$$

Equation 18 is valid whenever the quantity within the square brackets is non-negative. A negative value indicates that the motion was arrested during the second phase while the climber was building up to his maximum grip $G_{\max}$. Clearly, if the maximum drag resistance $4\mu_s G_{\max}$ is less than the climber s weight, the fall will never be arrested. The velocity of the climber, given by Eq. 13, can never go to zero.

Using a simple reaction time $t_r = 0.2$ seconds, the fall height H of Test Subject 1 may be calculated using Eq. 18. From Fig. 1 we obtain his weight, $W = 200$ lbs, his maximum grip, $G_{\max} = 119$ lbs, and his onset time, $t_0 = 0.25$ seconds. Table B-1 provides his coefficient of sliding friction, $\mu_s = 0.72$. Thus,

$$H = \frac{1}{2}(32.2)(0.2)^2 \left\{1 + \left(\frac{0.25}{0.20}\right)^2 \left(1 - \frac{4(0.72)119}{3 \cdot 200}\right) + 2\left(\frac{0.25}{0.20}\right) + \frac{\left[1 + \left(\frac{0.25}{0.20}\right)\left(1 - \frac{2(0.72)119}{200}\right)\right]^2}{\frac{4(0.72)119}{200} - 1}\right\} = 3.94\,ft. \qquad \text{Eq.19}$$

Using the data provided in Tables I and II the fall height was computed for each test subject and the results are tabulated in Table I. This table shows that four candidates cannot check their falls, two others fall 5 and 7 feet and the remainder fall less than 4 feet. The fall heights were also computed for each test subject when they wore nine different types of gloves. These findings are displayed in Table II where we observe that over 60% of the falls are critical (over seven feet).

There is an inverse relationship between the fall height H and the coefficient of sliding friction μ_s which is portrayed in Fig. 6 for test subject 2. This curve is asymptotic to the vertical line $\mu_s = W/(4G_{\max})$. Friction coefficients to the right of this line will cause test subject 2 to stop falling; values to the left of the line lead to unchecked falling.

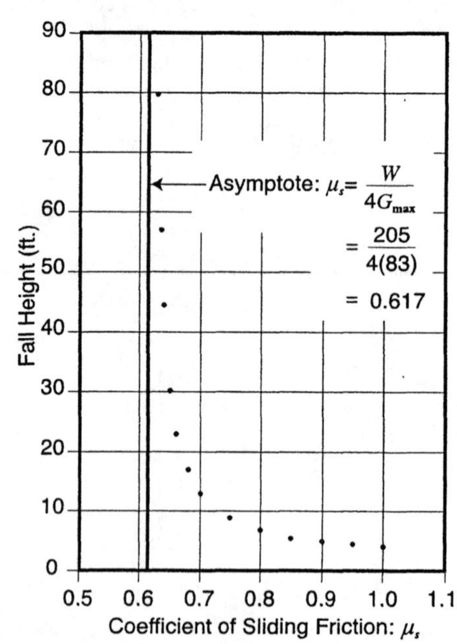

Figure 6 - *H* vs. μ_s: *Test Subject 2*

Table I - Fall Heights ($t_r=0.2$ seconds)

Test Subject	Onset Time t_0 (sec.)	Weight W (lbs)	Max.Grip $G_{\max}$ (lbs)	Sliding Friction μ_s	$\frac{4\mu_s G_{\max}}{W}$	Fall Height H (ft.)
1(M)	0.25	200	119	0.72	1.71	3.94 ft.
2(M)	0.226	205	83	0.63	1.02	79.23 ft.
3(M)	0.142	157	83	0.86	1.82	2.58 ft.
4(M)	0.107	250	113	0.70	1.27	4.91 ft.
5(M)	0.120	155	84	0.83	1.80	2.42 ft.
6(M)	0.140	126	72	0.75	1.71	2.77 ft.
7(M)	0.119	195	110	0.73	1.65	2.73 ft.
8(M)	0.070	184	102	0.65	1.44	2.89 ft.
9(M)	0.142	250	49	0.59	0.46	∞
10(M)	0.162	180	85	0.80	1.51	3.71 ft.
11(F)	0.160	150	31	0.67	0.55	∞
12(F)	0.144	140	52	0.82	1.22	6.61 ft.
13(F)	0.171	110	39	0.67	0.95	∞
14(F)	0.122	190	71	1.02	1.52	3.16 ft.

Table II - Fall Heights With Gloves (feet)

Test Subject	Glove Type								
	Felt Covered Rubber Gloves μ_s=0.43	Black Leather Gloves μ_s=0.45	Cotton Work Gloves With Leather Palms μ_s=0.54	All Cotton Gloves μ_s=0.60	Tan Leather Work Gloves μ_s=0.61	Excessively Worn Tan Leather Work Gloves μ_s=0.62	Canvas Gloves With Rubber Dotted Palms μ_s=0.66	Cloth Gardening Gloves With Rubber Dotted Palms μ_s=0.77	Knit Gloves With Rubber Cross Beaded Palms μ_s=0.89
1	74.29	25.56	7.56	5.55	5.34	5.15	4.55	3.59	3.04
2	∞	∞	∞	∞	∞	368.45	24.40	7.88	5.05
3	∞	∞	9.48	5.55	5.23	4.95	4.13	3.02	2.47
4	∞	∞	∞	13.22	11.07	9.57	6.37	3.65	2.71
5	∞	∞	7.45	4.68	4.44	4.23	3.59	2.68	2.22
6	∞	42.23	6.15	4.30	4.11	3.95	3.44	2.67	2.25
7	∞	71.54	6.02	4.12	3.94	3.78	3.27	2.52	2.12
8	∞	∞	5.39	3.57	3.40	3.25	2.80	2.13	1.79
9	∞	∞	∞	∞	∞	∞	∞	∞	∞
10	∞	∞	64.80	10.77	9.58	8.66	6.38	4.02	3.08
11	∞	∞	∞	∞	∞	∞	∞	∞	∞
12	∞	∞	∞	∞	∞	∞	∞	9.43	4.85
13	∞	∞	∞	∞	∞	∞	∞	15.53	6.27
14	∞	∞	∞	∞	∞	∞	∞	8.34	4.39

Rung Climbing Strategy

Climbing a fixed vertical ladder using a "rung grasping" strategy gives rise to a number of important observations:

1. A small horizontal inward acting force component must always be present to prevent a climber from rotating backward off of the ladder. This force is supplied by one or both hands.

2. To provide the required horizontal stabilization force, the climber's fingers need only maintain a "hook-like" geometry. No squeezing is necessary; however, it is acceptable.

3. Gravitational forces are resisted by the legs and by one or both hands. Force sharing among the appendages is constantly and abruptly changing which causes the climber to maintain a substantial "hook-like" grip but not a maximum power grip.

4. Loss of foothold causes a vertical downward translation of the climber without rotation away from the ladder. Both hands will always remain close to the plane containing the ladder rungs.

5. If there is a loss of foothold and the hands offer some counterforce to falling, the climber will fall a shorter distance than he would in free fall, i.e., less than 7.73 inches as calculated from Eq. 4 using $t_r = 0.2$ seconds.

6. At least one of the arms holding onto a rung will not be straightened out by a fall of only 7.73 inches. This implies that at least one hand will be protected by the shock attenuating property of a bent arm.

7. If the power grip used in the normal ascent or descent of a ladder is capable of supporting the weight of a climber for one second, the rung grasping strategy will protect the climber during a loss of foothold.

8. If the second hand is free at the time the foothold is lost, after the passage of 0.2 seconds it may be inserted four inches into the plane of the rungs in an attempt to grab a rung. Under ideal conditions, the hand can move inward at the hand speed constant, $v_h = 63$ inches/second [18-22]. The time required to reach four inches is,

   t = distance $\div v_h = 4/63 = 0.0635$ seconds

   At that time, the climber would have been falling for $t = 0.2 + 0.0635 = 0.2635$ seconds. The downward speed and descent of the free hand at $t = 0.2635$ seconds is given by Eqs. 5 and 4, respectively, as 8.48 ft/sec and 13.41 inches. The impact speed between the hand and a rung may be greater than 8.48 ft/sec; it cannot be lower. It is possible for the second hand to aid in a fall arrest scenario, but it cannot be relied on given such high closure speeds and the uncertainty of the rung catching dynamics.

9. If a fall is checked by one hand, the other may easily reach four or five rungs when attempting to reestablish a normal equilibrium configuration.

CONCLUSION AND OBSERVATIONS

1. For the siderail climbing strategy, all of the fall height calculations underestimate the actual fall heights that will be experienced during a loss of foothold. Some of the conservative assumptions reflected in the calculations include:

   a. The grip strength of the non-dominant hand was assumed to be equal to that of the dominant hand.

   b. The gripping surfaces of the ladder siderails were assumed to be as efficient as the grip provided by the Smedley Dynamometer.

   c. The grip/time relationships were assumed to follow the bilinear bounding curves.

   d. The climbing hand strength was taken as the non-fatigued strength measured in the laboratory.

   e. It was assumed that the maximum grip strength would not decrease when the hand slid down the siderail.

2. The fall heights calculated for bare hands are tabulated in Table I. Twenty-nine percent of the test subjects experienced uncontrolled falls; 7% suffered a critical fall (7 to 14 feet) and the remaining 64% were able to limit their falls to less than five feet. Thus, under unattainable, ideal conditions, a full 36% of the climbers would be seriously injured during a loss of foothold. Furthermore, it seems likely that in real world excursions, another 21% of the falls would be critical.

3. Loss of foothold calculations using gloved hands are exhibited in Table II. Fifty-two percent of the entries show uncontrolled falls; another 10% are critical. In real fall scenarios, another 12% of the entries will probably become critical.

4. The maximum grip strength of a gloved hand is less than that of a corresponding bare hand. Cochran [23] indicates a 7.3% decrease for cotton gloves and a 16.8% decrease for leather and cotton gloves. Accordingly, the fall heights displayed in Table II would be greater than shown since they were all calculated using grip tests conducted with bare hands.

5. The siderail grasping strategy does not deal effectively with a loss of foothold.

6. The rung grasping strategy will prevent climbers from falling after a loss of foothold if the power grip used in ordinary ascent and descent will resist their weight for about one second. The dangling arm and legs can reestablish a new purchase on the ladder in one second.

7. One bent arm will always be exerting an upward force on a rung during the onset of a fall. Downward motion is always restrained even during the simple reaction time; the power grip is always active; and the bent arm prevents impact loading of the hand.

8. It is assumed that the community of users of straight, extension and fixed ladders are capable of dangling from a rung with only one hand for one second.

9. The fourteen test subjects used in this study were arbitrarily selected; each could hang from a fixed ladder rung using one arm.

10. It should be noted that grip tests are never limited by fingers losing their grasp on the test device handle or trigger. The resistance of a "hook-like" finger geometry is always greater than the corresponding grip strength.

11. If a hand is not in contact with a rung at the onset of a fall, the ability to grab a rung is delayed by the time required to reach the rung and the simple reaction time. This delay time, in conjunction with a rung spacing of 12 inches, implies that the first rung encountered by a falling hand occurs after a drop of 13.41 to 25.41 inches. This corresponds to a hand/rung closure speed of 8.49 to 11.35 ft/sec. These speeds are 62% to 116% greater than the hand speed constant; they are too fast to be relied upon for fall protection.

12. To achieve their actual maximum grip strength, the average time required by the fourteen test subjects was 0.349 seconds with a standard deviation of 0.093 seconds. Their times ranged from 0.209 to 0.541 seconds. It should be noted that the test protocol required the candidates to squeeze the dynamometer as quickly as they could.

APPENDIX A

Fourteen candidates, ten males and four females, were tested to establish their grip/time relationships. Their ages ranged from 20 to 60 years; their occupations included: technician, mechanics, secretaries, engineers, librarians and college students. The testing program reflected the following protocol:

1. A Smedley type adjustable spring dynamometer, as illustrated in Fig. A-1, was used to measure grip strength in pounds. The grip size setting was maintained at two inches.

2. The dynamometer was mounted on a column with its handle in a vertical orientation to reflect the attitude of a fixed ladder siderail.

3. The dynamometer was mounted at the shoulder height of each candidate as depicted in Fig. A-2.

4. A digital camcorder was focused on the dynamometer dial. Frames were recorded at a time interval of 1/30th second throughout the test duration.

5. Candidates were instructed to begin the test by lightly grasping the dynamometer handle.

6. Candidates were told that they could commence squeezing the dynamometer at their discretion.

7. Candidates were asked to squeeze the dynamometer as fast as they could, as if startled.

8. The video recording was terminated when the candidate achieved his or her maximum grip.

9. Several days prior to the testing program, candidates were instructed in the operation of the dynamometer and allowed to practice with the instrument.

Figure 1 records the grip/time relationships of candidate1. The height, weight, shoulder height, gender and age of test subjects appears in the grip/time diagram.

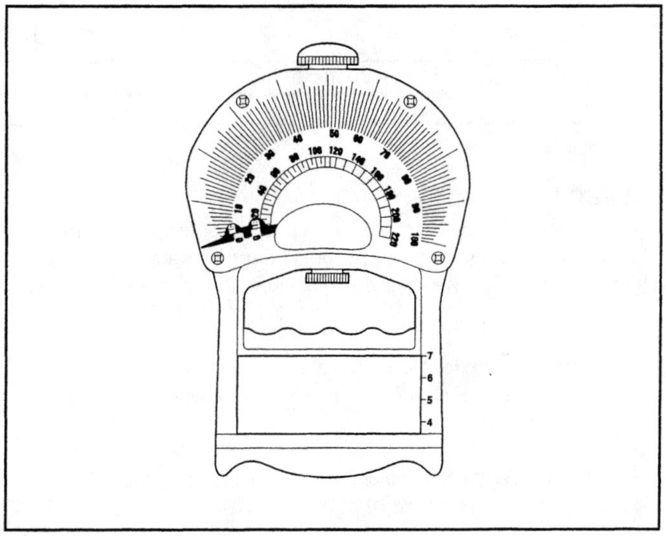

Figure A-1 - Smedley Type Adjustable Spring Dynamometer

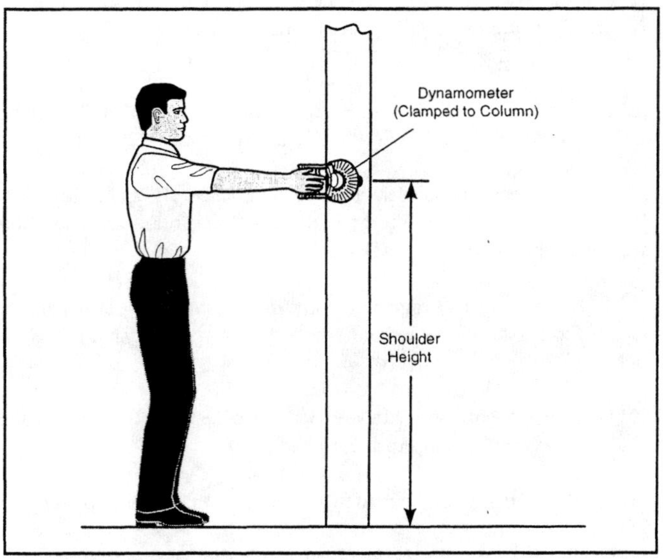

Figure A-2 - Grip/Time Test Setup

APPENDIX B

Table B-1 Coefficient of Sliding Friction - Test Subject 1

Station	Drag Force (lbs)	Coeff. Sliding Friction
1	4.0	0.70
2	4.0	0.70
3	4.2	0.74
4	3.9	0.68
5	4.1	0.72
6	4.3	0.75
7	4.2	0.74
	Average Coefficient:	0.72

REFERENCES

[1] Occupational Health and Safety Information, Safe Working Procedures. *Ladder Safety*. University College of the Cariboo, Kamloops, British Columbia, Canada. http://www.cariboo.bc.ca/hsafety/swp/ladder.htm (March 1, 2000).

[2] Canadian Centre for Occupational Health and Safety. *Important Safety Tips on Using Portable Ladders*. http://www.ccohs.ca/headlines/text31.html (March 1, 2000).

[3] Tapas, Tina D. "Take Ladder Safety One Rung at a Time," *Safety Focus Newsletter* (January/February 1998): 10-11.

[4] Labour Manitoba, Workplace Safety and Health Division. "Portable Ladder Safety," *Work Safe! Bulletin No. 177*. http://www.gov.mb.ca/labour/safety/publicat/bulletin/bltn177.html (March 1, 2000).

[5] Elsberry, Richard R. "The Ups and Downs of Portable Ladder Safety," *Electrical Apparatus* v. 53 #1 (January 2000): 48-49. *File 15: ABI/INFORM. DIALOG*. (February 29, 2000).

[6] Binsacca, Rich. "Ladder to Success," *Builder* v. 21 #14 (November 1998): 141. *File 15: ABI/INFORM. DIALOG*. (February 29, 2000).

[7] Office of Occupational Health and Safety. *Ladders & Step Ladder Safety*. University of Victoria, Victoria, British Columbia, Canada. http://web.uvic.ca/ohs/ladder.html (December 15, 1999).

[8] *Workplace Safety Program for Jay Bee Painting*. Jay Bee Painting, Ocala, Florida. http://www.atlantic.net/~titan/jaybee/safety.html (December 15, 1999).

[9] *Ladders*. Construction Safety Association of Ontario. Toronto, Ontario, Canada. Revised Edition, April 1984. p. 9.

[10] *Ladder Safety*. National Safety Council. Itasca, Illinois. pp. 8-9.

[11] Miller, Barrett. "Safe Ladder Management," *Professional Safety* v. 42 #11 (November 1997): 30-32.

[12] Hammer, Wilfried and Udo Schmalz. "Human Behaviour When Climbing Ladders with Varying Inclinations," *Safety Science* v. 15 (1992): 21-38.

[13] *Rincon Industries Ladder Safety Guidelines*. Rincon Industries, San Diego, California. http://www.rinconind.com/guide.html (March 1, 2000).

[14] Canada NewsWire. "Safety Alert - Don't Let Safety Get Rusty This Winter." November 22, 1999. *Dow Jones Interactive Publications Library*. (December 14, 1999).

[15] Brickman, Dennis B. "Ergonomic Studies of Grip Strength - Literature Review," *Triodyne Safety Brief* v. 6 #2 (June 1990).

[16] Barnett Ralph L, and Poczynok Peter J. "Ladder Rung v. Siderail Hand Grip Strategies," *Triodyne Safety Brief* v. 16 #4 (April 2000).

[17] McCormick, E. J. and M. S. Sanders. *Human Factors in Engineering and Design*. New York, McGraw-Hill, 5th ed., 1982, p. 198.

[18] Crawford, Alex. 1984. " 'Hand Speed Research Need Questioned,' in Power Presses Regulations are Thwarted." *Health & Safety at Work 6*, no.7 (March): 45.

[19] Horbury, C.R.J. et al. 1992. "European Standards for Machinery Safety" in *Contemporary Ergonomics*. Edited by E.J. Lovesey. Washington: Taylor & Francis.

[20] Knass, P. 1974. "Safety of Metalworking Press Operatives When Reaching Back into the Ram Zone of the Press During Ram Travel." Translated by the Health and Safety Executive Translation Services. July 1981. *Die Berufsgenossenschaft* (June): 239-246.

[21] "Mechanical Power Presses," *29 CFR 1910.217*. Washington: Occupational Safety and Health Administration, effective March 7, 1996 (as published in 29 CFR Parts 1900 to 1910.999, Revised as July 1, 1996, pp. 590-591).

[22] Roozbazar, A. 1977. "Ergonomics of Machine Guarding." *National Safety News* (July): 53-59.

[23] Cochran, David J., et al. "An Analysis of Grasp Force Degradation with Commercially Available Gloves," *Proceedings of the Human Factors Society, 30th Annual Meeting*, Dayton, OH, September 29 - October 3, 1986. Pp. 852-55.

Proceedings of
2001 ASME International Mechanical Engineering Congress and Exposition
November 11–16, 2001, New York, NY
SERA-Vol. 11

IMECE2001/SERA-24002

CHILDREN'S OUTERWEAR DRAWSTRINGS AND SLIDING BOARD SAFETY ANALYSIS

Dennis B. Brickman
Triodyne Inc.
5950 W. Touhy
Niles, Illinois 60714
(847) 677-4730
(847) 647-2047 (Fax)
brickman@triodyne.com (e-mail)

ABSTRACT

A two year old child strangled to death when the drawstrings of her jacket became lodged in a catch point hazard at the top of a residential sliding board. Approaches utilized in the safety analysis include accident reconstruction, safety literature review, standards research, an accident statistics survey, and an evaluation of alternative jacket and sliding board designs. Results of the analysis indicate there are technically and economically feasible design alternatives which prevent the child strangulation hazard associated with the jacket drawstrings and sliding board.

INTRODUCTION

A two year old child was strangled when the drawstrings of her jacket became lodged in a catch point at the top of a backyard slide. The child was last seen alive on the flat surface at the top of the slide and was found not breathing and with no pulse approximately 7 to 9 minutes later laying on the ground a foot in front of the bottom of the sliding board. There were no witnesses to the accident. After the accident, the coroner found a furrow most prominent at the posterior lateral aspect of each side of the child's neck and crossing the entire dorsal surface of the nèck. The coroner concluded that the cause of death was accidental drawstring ligature strangulation.

The subject jacket worn by the child at the time of the accident is depicted in Fig. 1. At the top of the neck of the jacket, there are two decorative drawstrings with round toggles at the ends. Each drawstring is approximately 22.9 cm (9 in.) long. The neck drawstrings serve no functional purpose. The jacket was manufactured in 1992. The jacket was purchased by the child's mother at a second hand store on 9/26/98. The accident occurred ten days later on 10/6/98.

The subject residential sliding board involved in the accident is shown in Fig. 2 and Fig. 3. Eight catch points for neck drawstrings have been identified at the top of the subject sliding board. The sliding board was manufactured in 1992 or 1993.

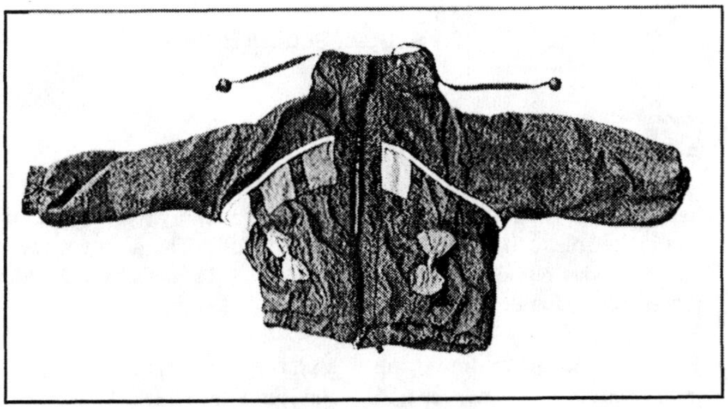

Fig. 1 - Subject Jacket

Fig. 2 - Subject Sliding Board

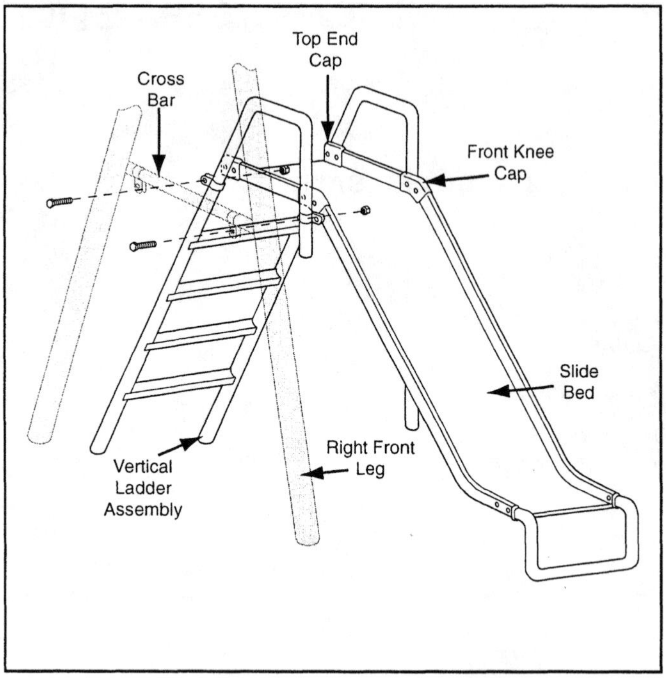

Fig. 3 - Subject Sliding Board

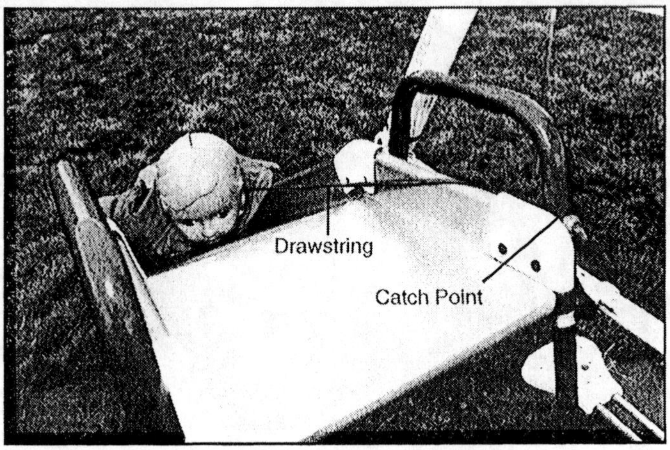

Fig. 4 - Drawstring in Catch Point

ACCIDENT RECONSTRUCTION

An accident reconstruction analysis was conducted using a 14.1 kg (31 lb), 78.7 cm (31 in.) anthropomorphic dummy wearing the subject jacket placed on the subject sliding board. The accident reconstruction was photographed and videotaped. Results of the accident reconstruction are as follows:

1. A single jacket drawstring was placed in the space between the interior of the front kneecap and the interior of the slide bed on both the left and right sides. During all attempts, the drawstring immediately pulled through the kneecap - slide bed interface.

2. A single jacket drawstring was placed in the catch point between the exterior of the right rear top end cap and the interior of the right rear vertical ladder assembly as shown in Fig. 4. The drawstring remained caught for more than 3 minutes and 57 seconds and then fully released by itself, allowing the anthropomorphic dummy to slide down the sliding board feet first on its front.

3. A jacket single drawstring was placed in the catch point between the right exterior of the horizontal portion of the slide bed and the cross bar. The drawstring remained in this catch point until it was manually released.

4. Both jacket drawstrings were placed in the catch point between the right front leg and the cross bar. After more than 50 seconds, both drawstrings fully self-released from the catch point, allowing the anthropomorphic dummy to slide down the sliding board feet first on its front.

5. During all the trials, the anthropomorphic dummy came to rest on the grass at the front of the slide after it slid down the sliding board feet first on its front.

6. When either a single drawstring or both drawstrings caught in a slide catch point, force was applied to the rear and sides of the anthropomorphic dummy's neck at the drawstring-neck interface. This applied force is consistent with the marks found on the child's neck by the coroner.

7. Potential causes for release of the drawstrings from the slide catch points at the time of the accident include the following:
   a. Geometry of the drawstring toggles with respect to the catch point.
   b. Movement of the jacket wearer while the drawstrings are caught.
   c. Finger aided release of the drawstring from the catch point by another child.

8. Times for full self-release of the drawstring from the slide catch point were achieved which are consistent with the length of time between when the child was last seen on the flat surface at the top of the sliding board and when the child was found laying on the ground in front of the bottom of the sliding board.

9. A tension test was conducted using the waist drawstring from the pants which accompanied the subject jacket. The failure load of this drawstring was 109 pounds, which is more than enough to suspend the anthropomorphic dummy's weight.

DRAWSTRING STRANGULATION HAZARD

Accident Statistics and Safety Literature

For over two decades, the drawstring strangulation hazard has been well documented by accident statistics and in the safety literature. On 5/7/81, the U.S. Consumer Product Safety Commission (CPSC) issued a Special Report on Accidental Strangulations (Ligature) of Children Less than 5 Years of Age [1]. According to this CPSC report, there were 20 strangulation incidents involving hoods or strings on clothing from 1973 through 1980. Furthermore, there were 13 fatal strangulations involving clothing or a rope catching on sliding boards. The CPSC reported that strings on the clothing of infants and toddlers appear subjectively to increase the probability of catching on something when compared to clothing without strings. It appears that a string on a piece of clothing is more likely to get caught on a small projection than is an entire hem of a garment.

In November 1995, the U.S. CPSC issued Guidelines for Drawstrings on Children's Outerwear to help prevent children from strangling or getting entangled by the neck and waist drawstrings of upper outerwear garments such as jackets and sweatshirts [2]. According to these CPSC guidelines, the CPSC received reports of 17 deaths and 42 non-fatal incidents involving the entanglement of children's clothing drawstrings from January 1985 through September 1995. Over two-thirds of the deaths and non-fatal incidents involved hood/neck drawstrings. The majority of these cases involved playground slides. Typically, as the child descended the slide, the toggle or knot on the drawstring got caught in a small space or gap at the top of the slide. Examples of catch points include a protruding bolt or a tiny space between the guardrail and the slide platform. As the child hung by the drawstring, suspended part way down the slide, the drawstring pulled the garment up taut to the neck, strangling the child. Victims in these cases ranged in age from 2 through 8 years old.

Prior to the manufacture of the subject jacket, the jacket manufacturer received notice of an accident involving one of its jackets. According to U.S. CPSC epidemiologic investigation report for case No. 900503CBB1401, in March of 1990, a 3-1/2 year old female had a jacket hood tied on her head with strings with plastic tabs at the ends. The victim was using a slide when one of the tabs caught in a catch on the slide causing the victim to hang on the incline with the string partially around her neck. A notification letter was sent to the jacket manufacturer by the U.S. CPSC for epidemiologic investigation report number 900503CBB1401 on 8/30/90.

Codes and Standards

Following three strangulation deaths of children from drawstrings and after voluntary efforts to eliminate strings were found to be unsuccessful, the United Kingdom issued The Children's Clothing (Hood Cords) Regulations in 1976 [3]. These mandatory regulations prohibited the sale of children's outer garments having a hood designed to be secured by means of a cord drawn through the material. According to a 5/5/94 CPSC memorandum which presents options to address risks to children from clothing string/cord entanglement, no deaths involving hood strings have been reported since the 1976 United Kingdom regulations came into operation [4].

On 1/10/96, a Provisional Safety Specification for Drawstrings on Children's Upper Outerwear was issued by The American Society for Testing and Materials (ASTM) [5]. The performance requirements of this ASTM provisional safety standard state that there shall be no drawstrings in the hood and neck area of children's upper outerwear sizes 2T to 12. This standard, ASTM F1816-97, was approved on 6/10/97 and published in August of 1998 [6].

DESIGN ALTERNATIVES

Warnings

In the fall of 1991, the CPSC released a Consumer Product Safety Alert entitled, Strings Can Strangle Children on Playground Equipment [7]. This safety alert contains a warning and pictograph as shown in Fig. 5. In April 1994, the CPSC Memorandum on Options to Address Clothing String Entanglement Hazards stated that the hazards of clothing strings are not apparent to most consumers [8]. The CPSC concluded that deaths will likely continue unless some action is taken to modify garments to eliminate or reduce the strangulation hazard. After the subject jacket was manufactured, the jacket manufacturer placed warning labels in subsequent jackets regarding the neck drawstring strangulation hazard.

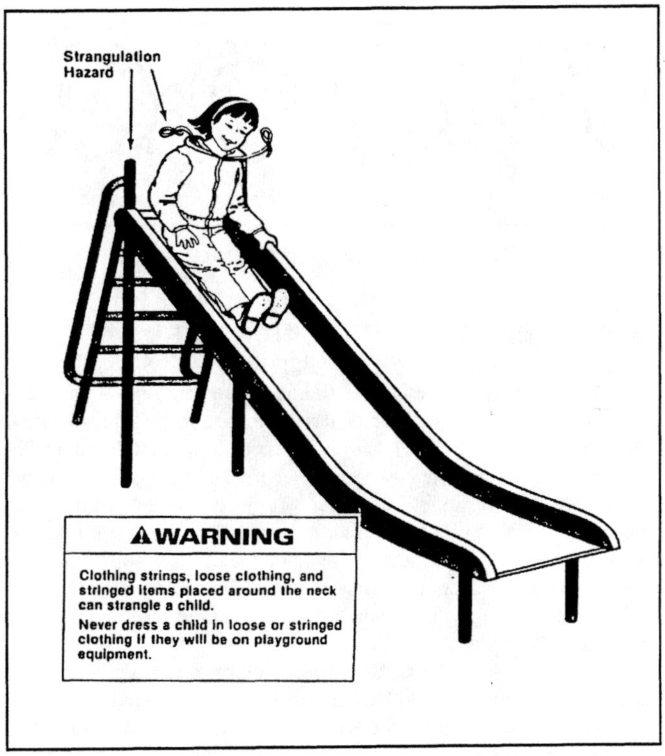

Fig. 5 - Consumer Product Safety Alert

Breakaway Strings

During a meeting between the CPSC staff and manufacturers of children's outerwear with strings on 4/22/94, the issue was raised about using a breakaway string or cord [9]. The CPSC staff stated that since a very small amount of pressure on the neck can slow or stop oxygen getting to the brain, a breakaway cord is probably not feasible. Also, it would break when a child pulled on it to tie it and would probably come apart in the laundry.

Eliminate Drawstrings

On 7/7/94, the CPSC issued a release stating that manufacturers and retailers had agreed to modify or eliminate drawstrings from hoods and necks of children's clothing [10]. On 9/28/95, the CPSC issued a Human Factors Analysis of Drawstrings on Children's Garments memorandum where the human factors staff concluded that strings at the neck of children's outerwear garments are a strangulation hazard and that no safe length of string can be recommended [11]. On 11/16/95, the CPSC took the additional step of issuing voluntary guidelines that advise manufacturers to eliminate drawstrings and to replace them with safer alternatives such as snaps, buttons, velcro and elastic as illustrated in Fig. 6 [12]. According to CPSC documents, these clothing modifications are relatively simple and would reduce the cost of manufacturing [8 - 9]. It should be noted that in 1994, subsequent to the manufacture of the subject jacket, the jacket manufacturer made a jacket with a similar collar and zipper and without a drawstring.

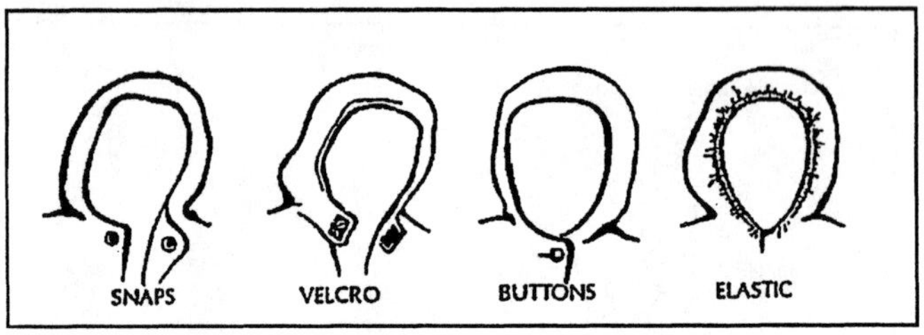

Fig. 6 - Drawstring Alternatives

SLIDING BOARD CATCH POINT HAZARD

Accident Statistics and Safety Literature

For over two decades, the sliding board catch point hazard has been well documented by accident statistics and in the safety literature. In March 1975, a CPSC Hazard Analysis of Injuries Relating to Playground Equipment cited two cases where girls, aged 2 and 4, were strangled when the poncho each was wearing caught on a vertical member of a slide railing [13]. In an August 1979 HIA Hazard Analysis Report on Public Playground Equipment, the CPSC reported that clothing entrapment was an important cause of sliding board related deaths where the clothing or rope caught at the top of the slide and the victim was strangled upon sliding [14].

In April 1990, the CPSC issued a report on Playground Equipment-Related Injuries and Deaths stating that it appeared that clothing was involved in at least 25 cases of strangulation from 1973 through September 1989, most often on slides and swings [15]. Items such as scarves, strings on clothing (mittens, jackets, ponchos, etc.), and hoods of jackets were reported to have been caught on protruding hardware (e.g., bolts), in narrow gaps between equipment components, on vertical posts, and on an open swing support hook.

According to CPSC accident statistics, the manufacturer of the subject slide has a history of strangulation accidents where children's clothing has caught on their sliding boards. These accidents occurred prior to the manufacture of the subject sliding board and/or prior to the subject accident.

1. According to U.S. CPSC IDI # 881024CCC2031, on 4/14/88, a 3 year old female died of asphyxiation when the string from her mittens became entangled on the slide rails of a playground slide. This caused the child's jacket to be pulled up around her neck.

2. According to the U.S. CPSC epidemiologic investigation report for case number 921026HCC1893, on 10/2/92, a string on a hooded sweatshirt worn by a 3 year old female became entangled on the handrail of a backyard sliding board. A notification letter was sent to the sliding board manufacturer by the U.S. CPSC on 3/31/93.

3. According to the U.S. CPSC epidemiologic investigation report for case number 940112CCC3183, on 7/15/93, two children aged 7 and 5-1/2 years became entrapped on separate occasions at the top of a backyard slide when a pull cord on their outer clothing got snagged in a space between two fabricated sheet metal sections.

4. According to the U.S. CPSC epidemiologic investigation report for Case No. 950127CBB1288, on 11/26/94, an 8 year old female became asphyxiated when the barrel clamp on the end of the waistband string of her coat became caught on the top edge of a backyard sliding board. As the victim went down the sliding board, the string became taut pulling the coat and zipper up around her neck. The victim never revived and died 10 days later as a result of brain damage.

Codes and Standards

The home playground equipment industry, as represented by the National Association of Children's Home Playground Manufacturers, Inc., developed Voluntary Product Standard PS 66-75, Safety Requirements for Home Playground Equipment, under the jurisdiction of the National Bureau of Standards [16]. This standard went into effect in 1976. The hazard of children's clothing or objects around a child's neck catching on a piece of equipment is not explicitly addressed in this safety standard.

A five year review and revision of Voluntary Product Standard PS 66-75 culminated in the publication of ASTM F1148-88, Standard Consumer Safety Performance Specification for Home Playground Equipment, in October 1988 [17]. According to the rationale for ASTM F1148-88, changes were made in order to minimize or eliminate the potential for clothing entanglement and falls from the top of slides. At the time of manufacture of the subject sliding board, the 1991 or 1993 revision of ASTM F1148 was in effect. In February 1992, the chair of the subcommittee of F-15 on consumer products that is responsible for the continued development of ASTM F1148 stated that items such as hoods, ponchos, scarves and mittens with strings connecting them can still become entangled on home playground equipment [18].

In November 1993, ASTM F1487-93, Standard Consumer Safety Performance Specification for Playground Equipment for Public Use, was published [19]. The introduction of ASTM F1487-93 states that the standard addresses fatalities reported to the CPSC resulting from entanglement of clothing or similar items on equipment. According to paragraph 6.3.2 of ASTM F1487-93, slides including protective barriers and their method of attachment and transition areas pose a greater risk of entanglement than other areas of play equipment.

DESIGN ALTERNATIVES

Warnings

According to the operating instructions requirements of Voluntary Product Standard PS 66-75, Safety Requirements for Home Playground Equipment, the operating instructions shall include statements warning the parent to dress children appropriately; examples would include wear well fitting shoes, and avoid ponchos, scarfs, and other loose-fitting clothing which is potentially hazardous while using equipment [16]. The operating instructions requirements were adopted by ASTM F1148 (1991 and 1993). The manual for the subject sliding board dated 3/3/93, contained a safe play tip which states, "Apparel such as hats with chin straps, helmet, and loose items around the waist, chest or neck can be hazardous when playing on this equipment." However, subsequently the sliding board manufacturer changed this warning in its manual to explicitly warn of the danger associated with drawstrings getting caught on playground equipment as follows: "TEACH YOUR CHILDREN THEY MUST: NOT wear loose clothing, clothing with hoods or drawstrings, jewelry, ponchos, scarfs or untied shoelaces or other loose-fitting clothing which is potentially hazardous while using equipment." According to the Safety Hierarchy [20], warnings represent the third priority in safety design.

Slide Guard

On 8/6/92, the CPSC issued a news release stating that the subject slide manufacturer was providing consumers with guards to prevent entanglement of strings on children's clothing in the handrail attachment area of certain slides. This news release refers to the report received by the slide manufacturer that a three year old girl strangled when her mitten strings, which were strung through her coat, caught at the point where the handrail attaches to the slide. The plastic guard offered by the slide manufacturer fills in the handrail to slide area where a child's clothing may catch and result in strangulation. It should be noted that applying safeguarding technology is considered to be the second priority in the Safety Hierarchy.

Eliminate Catch Point Hazard

According to the Safety Hierarchy, the first priority is to eliminate the danger through design by eliminating the hazard and/or the risk. In accordance with this first priority, the sliding board manufacturer has developed a design which eliminates several of the catch point hazards associated with the subject sliding board. Figure 7 depicts a sliding board with this alternative design which is located in the backyard immediately adjacent to the subject backyard.

Fig. 7 - Alternative Sliding Board Design

CONCLUSIONS

An accident reconstruction analysis utilizing the subject jacket, the subject sliding board, and an anthropomorphic dummy indicates that a strangulation hazard exists when the jacket drawstrings get caught in one of the sliding board catch points. Subsequently, the drawstrings fully release from the slide catch point and the anthropomorphic dummy comes to rest on the ground at the front of the sliding board after sliding down feet first on its front. The primary goal of this investigation is to make designers and users more aware of the drawstring-sliding board catch point strangulation hazard and to identify design alternatives to help reduce the number of these types of injuries. The Safety Hierarchy would suggest that attempting to eliminate the drawstring-sliding board catch point strangulation danger through design is the first priority. Applying safeguarding technology and warning of the danger are the second and third priorities respectively. Design alternatives such as snaps, buttons, velcro, and elastic for the subject jacket drawstrings and designs which eliminate catch points for the subject sliding board fall into the first priority of the Safety Hierarchy by eliminating the danger. These technically and economically feasible design alternatives which are available in the marketplace would prevent children from strangling due to drawstrings catching on sliding boards.

REFERENCES

[1] Rutherford, G.W. and Kelly, S., "Special Report: Accidental Strangulations (Ligature) of Children Less than 5 Years of Age," U.S. Consumer Product Safety Commission, Washington, DC, May 7, 1981.

[2] "Guidelines for Drawstrings on Children's Outerwear," *CPSC Document #1006*, U.S. Consumer Product Safety Commission, Washington, DC, Nov. 1995.

[3] "The Children's Clothing (Hood Cords) Regulations 1976," Her Majesty's Stationary Office, England, Jan. 2, 1976.

[4] Medford, R.L. and Tinsworth, D.K., "Options Paper: Strings on Children's Clothing," U.S. Consumer Product Safety Commission, Washington, DC, May 5, 1994, p. 5.

[5] "Provisional Safety Specification for Drawstrings on Children's Upper Outerwear," American Society for Testing and Materials, Philadelphia, PA, Jan. 10, 1996.

[6] "Standard Safety Specification for Drawstrings on Children's Upper Outerwear," *ASTM F1816-97*, American Society for Testing and Materials, Philadelphia, PA, Aug. 1988.

[7] "Strings Can Strangle Children on Playground Equipment," *Consumer Product Safety Alert*, U.S. Consumer Product Safety Commission, Washington, DC, 1991.

[8] Tinsworth, D.K., "Options to Address Clothing String Entanglement Hazards," U.S. Consumer Product Safety Commission, Washington, DC, Apr. 1994, p. v.

[9] "Meeting Between Consumer Product Safety Commission Staff and Manufacturers of Children's Outerwear with Strings," Apr. 22, 1994.

[10] Tyrell, E. "CPSC Works With Industry to Remove Drawstring Hazard," *Release #94-103*, U.S. Consumer Product Safety Commission, Washington, DC, Jul. 7, 1994.

[11] Trainor, C.M., "Human Factors Analysis of Drawstrings on Children's Garments," U.S. Consumer Product Safety Commission, Washington, DC, Sep. 28, 1995, p. 3.

[12] Begala, K., "Child's Death Spurs Effort to Release Guidelines for Drawstrings on Children's Jackets and Sweatshirts," *Release # 96-023*, U.S. Consumer Product Safety Commission, Washington, DC, Nov. 16, 1995.

[13] "Hazard Analysis of Injuries Relating to Playground Equipment," *NIIC-1201-75-H005*, U.S. Consumer Product Safety Commission, Washington, DC, Mar. 1975, p. 19.

[14] Rutherford, G.W., "HIA Hazard Analysis: Injuries Associated with Public Playground Equipment," U.S. Consumer Product Safety Commission, Washington, DC, Aug. 1979, p. 12.

[15] Tinsworth, D.K. and Kramer, J.T., "Playground Equipment-Related Injuries and Deaths," U.S. Consumer Product Safety Commission, Washington, DC, Apr. 1990, p. 21.

[16] "Safety Requirements for Home Playground Equipment," *Voluntary Product Standard PS 66-75*, National Bureau of Standards, Jul. 1976.

[17] "Standard Consumer Safety Performance Specification for Home Playground Equipment," *ASTM F1148-88*, American Society for Testing and Materials, Philadelphia, PA, Oct. 1988.

[18] Hendy, T.B., "Backyard Swingset Safety," *ASTM Standardization News*, Vol. 20, No. 2, Feb. 1992, p. 30.

[19] "Standard Consumer Safety Performance Specification for Playground Equipment for Public Use," *ASTM F1487-93*, American Society for Testing and Materials, Philadelphia, PA , Nov. 1993.

[20] Barnett, R.L. and Brickman, D.B., "Safety Hierarchy," *Journal of Safety Research*, Vol. 17, No. 2, Summer 1986, pp. 49-55.

Proceedings of
2001 ASME International Mechanical Engineering Congress and Exposition
November 11-16, 2001, New York, NY
SERA-Vol. 11

IMECE2001/SERA-24003

USE OF QUALITATIVE RISK ANALYSIS TO PRIORITIZE PROJECT RISK REDUCTION EFFORTS

Pamelyn G. Lindsey and Charlie A. Soczek
E. I. du Pont de Nemours & Co., Inc.
DuPont Engineering Technology - Process Safety & Fire Protection
1007 Market Street
Wilmington, DE 19898
Email: Pamelyn.G.Lindsey@usa.dupont.com and Charles.A.Soczek@usa.dupont.com

ABSTRACT

The risk associated with a highly toxic bulk storage facility was evaluated using a DuPont Qualitative Risk Assessment Protocol. The protocol used for the evaluation will be presented along with the results of the assessment. By application of this method a capital savings opportunity of over $500,000 can be realized as compared to the initial project scope estimate. These potential savings result from implementation of risk reduction measures based on an evaluation of both acute and chronic health effects on workers and the community. Additional process hazards analyses will be conducted to ensure adequate risk mitigation is implemented as the project design is finalized.

INTRODUCTION

Sound risk analysis can improve process safety and still help to reduce project cost. Rigorous methodologies can be employed to assess risk via qualitative and/or quantitative approaches. This paper will describe the application of a qualitative risk assessment protocol. This protocol was applied to a project to help prioritize risk reduction efforts.

The qualitative risk analysis methods used in this example estimate the risk associated with acute releases of a highly toxic material. Quantitative methods were used to estimate the consequences of worse case and typical releases. Chronic health effects on workers and the community were evaluated as part of the development of overall risk reduction strategies.

DESCRIPTION OF THE PROBLEM

The plant used in this example is located in a small to medium sized town. It is bordered on one side by a river, and by commercial and residential areas on the other sides. In order to provide a key ingredient to one of the plant's processes increased storage capacity for this toxic material is required. This raw material is shipped to the plant and must be off-loaded into a storage facility. Several options were evaluated from a safety, risk and economic perspective. A large storage facility was available on the site and was to be evaluated to determine if it was suitable for use. The tank had previously been used to store a non-toxic, non-corrosive material. Therefore, the tank was expected to be sound. This assumption needed to be verified with vessel mechanical integrity inspections and testing before the tank would be placed into service. The issue was – Is it safe to store such a large quantity of this material in the location? Ultimately – What is the risk associated with this storage option?

Like most efforts today, this was a fast track project. The plant needed this raw material for the next campaign. The safety aspects needed to be evaluated in a rigorous and timely manner.

OBJECTIVE OF THE ANALYSIS

Therefore, the intent of this assessment was to (1) analyze the risk associated with storage of this highly toxic material, (2) analyze the risk in a time frame that fits with the project objectives, and (3) prioritize risk reduction efforts.

OUTLINE OF QUALITATIVE RISK ANALYSIS PROTOCOL

Risk is a measure of human injury or economic loss, based upon both the likelihood and the magnitude of the loss or injury. It can be defined as the product of the expected frequency (events/unit time) and the consequences (effects/event) of a single hazardous event or group of events. Numerically the risk would then be expressed as effects/unit time. "For example, if analysis of an initiating event indicates that the event could result in two serious injuries every 150 years, the risk is calculated to be 2/150, or 1.3×10^{-2} serious injuries per year" [1].

Risk can be calculated via qualitative or quantitative methods. Qualitative risk assessments (qRA) use experience and engineering judgment as the basis for determining event consequences and frequencies. This is done as a result of the collaborative efforts of a multidiscipline team. The advantage of a qRA is that it can be less time and resource consuming than quantitative approaches. This method is also less narrowly focused than quantitative methods, allowing a broad spectrum of hazards to be evaluated.

Quantitative risk assessments (QRA) use mathematical models to develop numerical estimates of the expected frequency and consequences of potential hazardous events. The cost of conducting a QRA can be substantial; it is typically justified only when the risk reduction cost is high relative to the investment. Typically a qRA must precede a QRA to insure that hazards of significance have been properly identified. *Evaluating Process Safety in the Chemical Industry, A User's Guide to Quantitative Risk Analysis* provides guidance on when to proceed to a QRA [2]. *Guidelines for Chemical Process Quantitative Risk Analysis* provides details on how to conduct such analyses [3].

Most analyses are somewhere in between; that is, they are semi-quantitative. This was the approach used for the problem described here. As a result, the project saved an estimated $30,000 in analysis cost.

Prior to starting the qRA, screening Process Hazards Analyses had been conducted. The plant had identified several hazardous event scenarios including the worse case event(s) and the more typical events. A quantitative consequence analysis had been conducted for input into the qRA to ensure that the project team had an adequate understanding of the consequences of events.

The Qualitative Risk Assessment Protocol was then applied to analyze the risk [4]. In general, the consequence category is selected first (based in this example on the quantitative consequence analysis). Then, the frequency of failure of systems already in place to prevent the undesired event is selected. This is based on plant experience and engineering judgment. Hence the qualitative nature of the analysis. The combination of consequence and frequency is then used to determine the risk score, which in turn, is used to prioritize the risk reduction efforts.

Qualitative Risk Assessment Protocol

Step 1. Identify the event and the consequences qualitatively or quantitatively.
Determine the consequence category using Table 1.

Step 2. Identify the ways the event might occur. This is normally done through use of a process hazards analysis (PHA). Various PHA methodologies are described in *Guidelines for Hazards Evaluation Procedures*, Second Edition. [5].

Step 3. Determine the frequency category of the event using Table 2.

Step 4. The consequence category number and the frequency category number are combined to obtain a risk score from Table 3. This table was adapted from Reference [1].

Step 5. The risk score may be interpreted using the Table 4. This is an example of risk ranking categories given by Reference [1]. Each user must develop individual risk ranking criteria that are appropriate for the particular use.

Table 1. Consequence Category Matrix

Fire, toxic release, explosion, irreversible health effects:

Type of event/impact	Consequence Category C-1: minor	Consequence Category C-2: moderate	Consequence Category C-3: major	Consequence category C-4: catastrophic
Public safety and health	No injury or health effects	Minor injury or reversible health effects	Injury or moderate health effects. Liability for injury + damages $500M-$1MM	Death or irreversible health effects. Liability for injury + damages more than $1MM
Employee Safety and health	No injury or health effects	Minor (MTC) injury or reversible health effects	Multiple MTC injuries; 1-2 RWC/LWCs or moderate health effects.	One or more fatalities or irreversible health effects.

Source: DuPont Company, Qualitative Risk Assessment Protocol [4]

Table 2. Frequency Category Matrix

Category	Typical description	Approximate corresponding quantitative frequency (per year)
F-1 Extremely unlikely (or remote)	Not realistically expected to occur	$<10^{-4}$
F-2 Very unlikely	Not expected to occur, but not incredible	10^{-3} to 10^{-4}
F-3 Unlikely	Unlikely to occur in the plant's lifetime but could occur in one of a number of similar plants	10^{-2} to 10^{-3}
F-4 Likely	May occur at least once in the lifetime of the installation	$>10^{-2}$

Source: DuPont Company, Qualitative Risk Assessment Protocol [4]

Table 3. Risk evaluation matrix (adapted and used by permission from CCPS, Ref. 1). Source: JBF Associates, Inc., Knoxville, TN

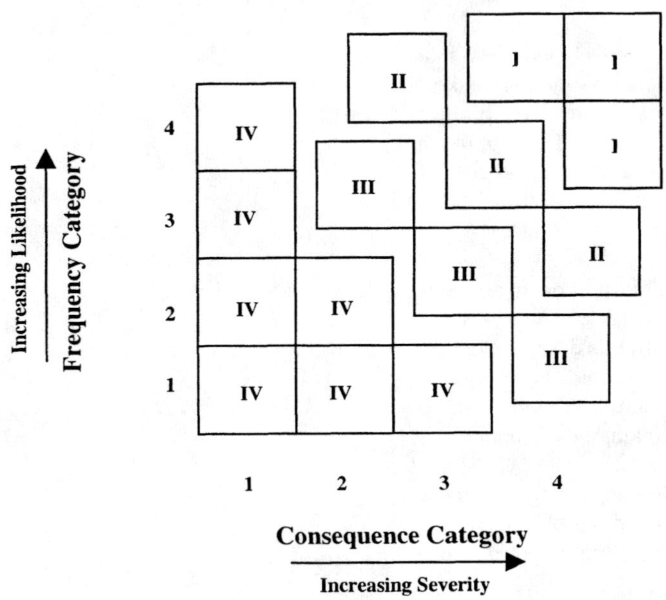

Table 4. Interpretation of the risk score (adapted and used by permission from CCPS, Ref. 1)

Risk Score	Description	Action required	PHA recommendation
I	Intolerable	Should be mitigated with engineering and/or administrative controls to a risk ranking of III or less within a specified time period, such as six months.	Yes
II	Undesirable	Should be mitigated with engineering and/or administrative controls to a risk ranking of III or less within a specified time period, such as 12 months.	Yes
III	Tolerable with controls	Should verify that procedures and controls are in place, establish emphasis that they are maintained.	Case-by case evaluation. Evaluate adequacy of existing controls.
IV	Tolerable as is	No mitigation required.	No. Consider an improvement opportunity (outside the PHA), if appropriate.

ADDITIONAL GUIDANCE

Tables 5 and 6 may be used for additional guidance on the selection of the frequency and consequence categories. Table 6 would be of particular use for qualitative evaluations of the consequences of events, when quantitative results are not available. The frequency and consequences should be evaluated specifically for the event in question, and not in a global sense for the entire area (plant). It is possible to have a low risk score for one event and a high risk score for another event.

There are three approaches to applying the consequence category matrix or the frequency category matrix. Any may be appropriate and the team should use its judgment to select the most appropriate approach for the specific situation.

- "Weak Link Approach"—This approach assumes that the overall safety chain is only as strong as its weakest link. Hence, the worst individual score (i.e., highest number) is used in the risk evaluation. This approach should be used when the team believes that the risk of the event in question could happen because of a failure of any of the engineering and management systems.

- "Target Category Approach"—This approach assumes that only some (or one) of the generic categories are directly applicable to the hazard. In this case, the lowest individual score among only the "target" categories is chosen. This approach should be used when some categories are much more important than the others:

  - Example 1: For a rail-car unloading operation where the hazard is accidental unloading of the wrong material; there may be few engineering controls, and safety is dependent upon the training and operation discipline of the unloading person. The event frequency evaluation should emphasize the human factors categories.

  - Example 2: For a highly automated, heavily interlocked system that has little operator interaction, the engineering controls categories (e.g., testing, reliability) are likely most important.

- "Balanced Approach"—The analyst simply averages the scores and rounds up or down to get an overall score. This can be used when the team feels there is some redundancy in the overall lines of defense. For example, a failure of an interlock is likely to be corrected through strong operator training and/or management systems.

Table 5. Supplemental Event Frequency Evaluation Matrix

Attribute	Frequency category F-1: extremely unlikely	Frequency category F-2: very unlikely	Frequency category F-3: unlikely	Frequency category F-4: likely
Engineering controls				
Lines of defense[a]	Two or more passive systems, independent, no reliability issue.	Two or more, at least one passive and reliable.	One or two, complex, active. Some reliability issues may have common mode weaknesses.	None or one, complex, active. Poor reliability.
Testing (interlocks, mechanical integ. and emergency systems)	Well-documented test protocol. Complete function check. Good results. Rare failures.	Regular tests, function check may be incomplete. Problems are uncommon.	Not checked often. History of trouble. Some tests claimed, not done.	Underlined, unchecked or unappreciated.
Incident history	No major events, very few minor events. Prompt, corrective actions taken.	No major incidents. Perhaps minor incidents. Causes understood and learning is captured.	One major incident. Causes not completely understood. Questions remain if corrective action is adequate.	Many incidents. Near misses. Failure to learn.
Operating experience	Process well under-stood. Rare upsets are acted upon promptly.	Rare upsets. Most causes understood. Effective corrective action.	Chronic minor upsets not all explained or acted upon. More serious ones are flagged and eventually resolved.	Routine upsets, many never explained. Excursion common and causes not well understood.
Rate of change	Stable process; potential hazards are understood Data always available to support SOL and SOCs.	Reasonable rate of change. May be new technology, with some uncertainty. Good PHAs.	Rapid changes or new technology. Fair PHAs, not always given deep thought. Operational limits uncertain.	Rapid changes. New technology. Incomplete or poor PHAs. Learning as you go.
Human factors				
Training and procedures	Clear, unambiguous OD[b]. Discipline in place to follow. Errors flagged and corrected immediately. Routine refresher training, including normal, nonroutine and emergency proce – dures. All contingencies covered.	Critical OD in good shape. Others have nonfatal errors or weaknesses. Routine audits and reviews. Personnel are familiar with procedures.	OD exist. Not updated regularly or token review. Poor emergency training procedures.	Unaware of OD. Training by work of mouth. "Black books." Excess verbal directions. Ad hoc operation. No emergency training.
Skill and performance (operators, mechanics, supervisors, contractors)	Multiple experienced operators on all shifts. No significant overwork or boredom. Optimal stress level. All are well qualified. Dedication evident. People care. Hazards clearly understood and appreciated.	Some new people. Never all on a single shift. Occasional brief fatigue. Some boredom. People know what they are qualified to do and their limitation. Healthy respect for hazards.	Possible shift of all rookies, but not common. Periods of group fatigue for short periods, extensive boredom. People not expected to think. People may assume more than they know. Not everyone under-stands the hazards.	High turnover. One or more shifts with no experienced people. Excess overtime, fatigue common. Disruptive work schedule. Poor morale. Jobs performed by people with question-able skills. No definition of job limitations. Unaware of hazards.

Source: DuPont Company, Qualitative Risk Assessment Protocol [4]

[a]Definitions for lines of defense:
Passive: Requires no human intervention or motive power sources
Active: Electrical interlock system or requires human action

[b]Operating directions

Table 6. Supplemental Consequence Evaluation Matrix

Type of event/impact	Consequence category C-1:minor	Consequence Category C-2: Moderate	Consequence category C-3: major	Consequence category C-4: catastrophic
Financial/Investment effects				
Production loss/damage	Less than one-week shutdown. Damage + prod. loss less than \$0.5 million.	1 to 2-week shut-down. Damage + prod. loss \$0.5-1 million.	2 to 4 – week shutdown. Damage + prod. loss \$2-3 million	Longer than one-month down, loss >\$3 million.
Building damage	Building performs function, reusable with minor repairs. Little risk to occupants.	Building performs function, but is not reusable without major repair close to or exceeding replacement cost. Risk of some injury to building occupants.	Loss of building integrity. Probably severe injury to occupants. Possibility of fatal injuries to some.	Building demolished. Fatal injury to occupants is likely.
Environnemental effects				
On-site effects	On-site nuisance: continuous noise, fallout on cars, etc.	Severe nuisance: intense smoke, significant dust or smell, etc.	On-site concern: flames, shock wave, toxic release	On-site major concern: significant fire or explosion, site evacuation
Off-site effects	Short duration flaring, unusual noise, smoke, or smell causing a few complaints from neighbors.	Continuous flare or noise, unpleasant and persistent smell, dust, or fallout.	Fire and smoke affecting off-site areas. Explosion shock felt off-site. Major dust or fallout.	Major gas or smoke discharge, large off-site evacuation. Significant release of carcinogen. Major damage to flora/ fauna or repeated major events.
Liquid Discharge	Spill all or largely contained. Discharge greater than report-able amount.	Significant loss to river or water. Local effect only. Low fish kill potential.	Large amount lost to water. Serious effects on river, significant fish kill. Failure to meet permit require-ments. Reaction by environmental groups.	Major loss to river, major recovery effort. Serious effect on wildlife. Permanent/continu-ous soil or water contamination.
Legal impact	Reportable incident.	"Warning" from government agency.	Significant fine.	Major fine, criminal prosecution.
Media attention	Minor publicity; local TV/press.	Significant local publicity; interviews.	Considerable local publicity; some national publicity.	"Headline" locally and nationally. Public outcry.

Source: DuPont Company,
Qualitative Risk Assessment Protocol [4]

APPLICATION OF PROTOCOL TO THE PROBLEM

As stated above, quantitative consequence analysis had previously been conducted to ensure that the project team had an adequate understanding of the consequences of events. The methods used in analyzing the consequences are based on the DuPont Engineering Consequence Assessment Technical Guidance Manual [6]. These methods are very prescriptive and designed to produce conservative results. *Guidelines for Use of Vapor Cloud Dispersion Models* provide an overview of modeling procedures [7].

Two events were evaluated, (1) a large/worse case release from the vessel and (2) a more typical small release. The large release was modeled as a 2" hole in the vessel and the small release was modeled as a ¼" hole in the vessel.

Consequence modeling of the downwind effects of these releases was conducted using the PHAST® model developed by Det Norske Veritas (DNV). This model is a comprehensive consequence analysis tool and uses the same Unified Dispersion Model (UDM) as a larger quantitative risk assessment package, SAFETI®, also developed by DNV. This is of particular use when transitioning from a consequence analysis to a QRA.

A range of meteorological conditions was evaluated, including very stable, low wind speed conditions (F stability and 1.5 m/s wind speed) and neutral stability with moderate wind conditions (D stability and 5 m/s wind speed). The surface roughness values used in the meteorological data are typical of urban areas. All other atmospheric conditions were typical for the site being studied.

As recommended under the provisions of the United Stated Environmental Protection Agency Risk Management Plan, Emergency Response Planning Guidelines (ERPGs) were used for toxic endpoint values. ERPGs are developed by a committee of the American Industrial Hygiene Association [8]. These values are intended to provide an estimate of the concentration ranges where one might reasonably anticipate observing adverse effects as described in the definitions for ERPG-1, ERPG-2, and EPPG-3. These values were available for the chemical being studied and are defined as follows.

> ERPG-1 is defined as the maximum airborne concentration below which it is believed that nearly all individuals could be exposed for up to 1 hour without experiencing other than mild transient adverse health effects or perceiving a clearly defined objectionable odor. The ERPG-1 value for this chemical is 10 ppm. This value was not of particular importance to the current study.
>
> ERPG-2 is defined as the maximum airborne concentration below which it is believed that nearly all individuals could be exposed for up to 1 hour without experiencing or developing irreversible health effects or other serious health effects or symptoms that could impair an individual's ability to take protective action. The ERPG-2 value for this chemical is 50 ppm.
>
> ERPG-3 is defined as the maximum airborne concentration below which it is believed that nearly all individuals could be exposed for up to 1 hour without experiencing or developing life-threatening health effects. The ERPG-3 value for this chemical is 500 ppm.

RESULTS OF ANALYSIS

Based on dispersion analysis conducted using the PHAST@ program, the following consequence results were obtained for the events evaluated.

The downwind distance to the worse case release resulting from a 2" hole in the vessel at F 1.5 m/s is:

> ERPG-3 effects - up to 2,700 feet downwind
> ERPG-2 effects - up to 11,200 feet downwind

The downwind distance resulting from a more typical small flow release resulting from a ¼" hole in the vessel at F 1.5 m/s is:

> ERPG-3 effects - up to 320 feet downwind
> ERPG-2 effects - up to 1,620 feet downwind

Figures 1 – 4 show a graphical representation of these results and also show the results for the more typical day conditions. For simplicity the qRA was conducted using the consequence analysis results of the F Stability, 1.5 m/s cases. This wind condition typically gives the worst case consequences. This was the case for this example.

Figure 1. 2" Hole in Vessel, Typical Night Conditions, F Stability, 1.5 m/s

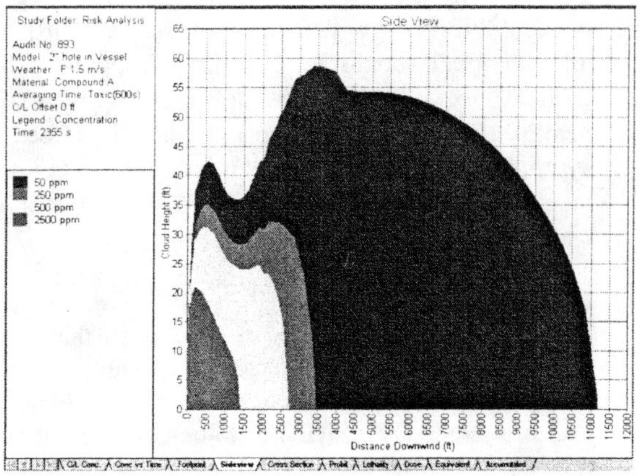

Figure 2. 2" Hole in Vessel, Typical Day Conditions, D Stability, 5 m/s

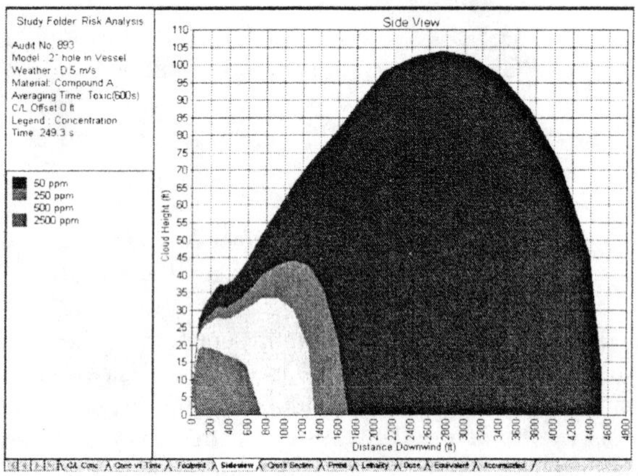

Figure 3. ¼" Hole in Vessel, Typical Night Conditions, F Stability, 1.5 m/s

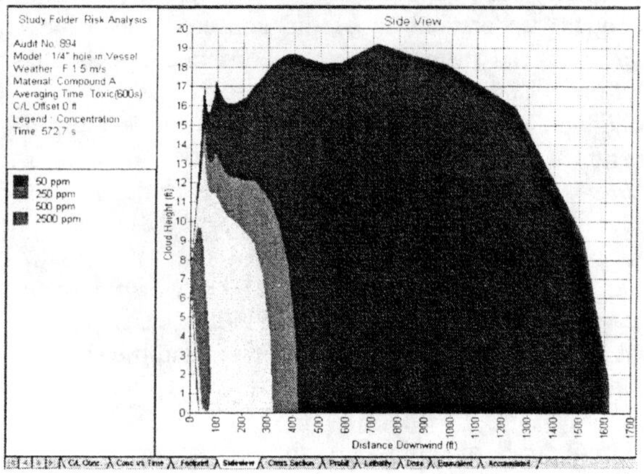

Figure 4. ¼" Hole in Vessel, Typical Day Conditions, D Stability, 5 m/s

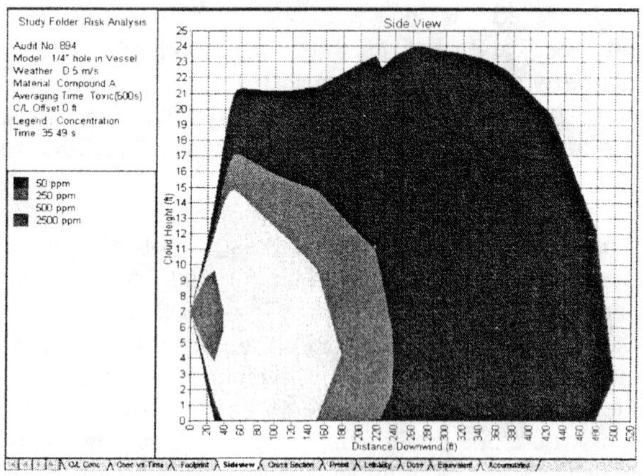

The "Worst Case" – 2 inch Hole in Vessel

The distance to the nearest off-site population is greater that 0.5 mile and is in the direction of the prevailing wind. Based on the dispersion results it is possible to have ERPG-3 concentrations off-site if this event occurred.

Following the protocol as outlined in Step 1, Table 1 was used to determine the consequence category. It was determined that the consequences are "C-4: Catastrophic". This is based on the public safety and health impact definition that the event could result in "death or irreversible health effects, liability for injury, and damages more than $1MM, as well as, the employee safety and health definition that the event could result in "one or more fatalities or irreversible health effects".

Following the protocol for Step 3, Table 2 was used to determine the frequency category. It was determined that this event is "extremely unlikely or remote" as based on experience from previous analysis and plant experience. The frequency category is then "F-1".

Combining the consequences and frequency as outlined in Step 4 results in a Risk Score of "III" from Table 3. Interpretation of the risk score determined that the event is "Tolerable with Controls" based on Table 4. The project "should verify that procedures and controls are in place and establish emphasis that they are maintained". PHA recommendations should be evaluated on a case-by-case basis as well as the adequacy of existing controls. For example, it was suggested that proper (or upgraded) bolting be used and that vessel/piping inspections be utilized to verify that the equipment is suited for its purpose.

The "Small Typical Case" - ¼ inch Hole in Vessel

Distance to ERPG-2 for the worse case weather condition is 1,580 feet. The distance to ERPG-3 for the worse case weather condition is 430 feet. The distance to the nearest off-site population is greater than 0.5 mile and is in the direction of the prevailing wind. Based on these results this event would not have off-site impact.

Following the protocol as outlined in Step 1, Table 1 was used to determine the consequence category. It was determined that the consequences are "C-2: Moderate". This is based on the public safety and health impact definition that the event could result in "minor injury or reversible health effects, as well as the employee safety and health definition that the event could result in a minor medical treatment case (MTC) injury or reversible health effects".

Following the protocol for Step 3, Table 2 was used to determine the frequency category. It was determined that this event is "likely" based on plant experience. The frequency category is then "F-4".

Combining the consequences and frequency as outlined in Step 4 results in a Risk Score of "II" from Table 3. Interpretation of the risk score determined that the event is "Undesirable" from Table 4. The risk "should be mitigated with engineering and/or administrative controls to a risk ranking of III or more". PHA recommendations should address the risk within the project scope. For example it was suggested that proper predictive and preventive maintenance procedures be implemented for hoses, piping, vessels, etc. Proper air monitoring should also be implemented.

NEXT STEPS

These results were then applied to the project. Several adjustments were made to the project scope. This resulted in potential reductions in the project cost of $500,000. These potential reductions resulted from proper application of risk reduction measures. These changes made sure that efforts were directed to activities that would actually reduce risk rather than just the perception of reduction. For example, the project team was advised not to weld flanges shut in response to the potential for a worse case event since the frequency of such an event was extremely unlikely. The project team was advised rather to direct risk reduction effort to the more likely events that would affect plant personnel. Industrial hygiene monitoring is one way to detect such events before a large problem develops.

Additional process hazards analysis shall be conducted on the project per Corporate Project Implementation Guidelines to ensure that these and other risk mitigation are put in place.

This qRA provided the project team with invaluable information and speed to allow the project to proceed to meet business objective.

Disclaimer - This protocol is for use as a qualitative risk assessment tool. It does not replace the need for a formal quantitative risk assessment when circumstances warrant. Be especially wary of using this protocol when the potential consequences of a particular event are extremely high or catastrophic. Judgment must be applied to the use of the risk score. Each user must establish individual risk management criteria when interpreting the risk score and risk reduction efforts that are required.

ACKNOWLEDGEMENTS
The authors would like to acknowledge the efforts and cooperation of the qRA and project team as resourced by Dick Hawthorne and the plant personnel as led by Mark Casey. The authors would also like to thank the Corporate Process Safety Management group as led by Art Burk, the PHA Leadership Team Chairman/author of the protocol, Harry Glidden and the business Safety Health and Environment lead, Rob Pinchot. Finally, the authors would like to thank all the reviewers and contributors to this manuscript.

REFERENCES
1. Center for Chemical Process Safety. *Guidelines for Evaluating Process Plant Buildings for External Explosions and Fires.* New York: American Institute of Chemical Engineers. 1996.
2. American Chemistry Council and Center for Chemical Process Safety. *Evaluating Process Safety in the Chemical Industry, A User's Guide to Quantitative Risk Analysis.* New York: American Institute of Chemical Engineers. 2000.
3. Center for Chemical Process Safety. *Guidelines for Chemical Process Quantitative Risk Analysis.* New York: American Institute of Chemical Engineers. 1989.
4. DuPont Company, Qualitative Risk Assessment Protocol. Appendix to Process Hazards Analysis Standard. [Unpublished Information]. E. I. Du Pont de Nemours & Co., Wilmington, DE. 1998.
5. Center for Chemical Process Safety. *Guidelines for Hazards Evaluation Procedures.* Second Edition. New York: American Institute of Chemical Engineers. 1992.
6. DuPont Company, Consequence Assessment Technical Guidance Manual. [Unpublished Information]. E. I. Du Pont de Nemours & Co., Wilmington, DE. 1997.
7. Center for Chemical Process Safety. *Guidelines for use of Vapor Cloud Dispersion Models.* New York: American Institute of Chemical Engineers. 1987.
8. American Industrial Hygiene Association Press. *Emergency Response Planning Guidelines.* Virginia: American Industrial Hygiene Association. 2000

Proceedings of
2001 ASME International Mechanical Engineering Congress and Exposition
November 11-16, 2001, New York, NY
SERA-Vol. 11

IMECE2001/SERA-24004

THE MACHINE OPERATOR'S JAMMED-FEEDSTOCK-CLEARING TASK: A SAFETY DESIGN CHALLENGE

J.R. Etherton and E.A. McKenzie, Jr.
National Institute for Occupational Safety and Health, Morgantown, WV

ABSTRACT

Workers are often exposed to injury risk from sudden and unexpected machine movement when clearing jammed material from process machinery. An engineered safety response can effectively intervene to control or eliminate this risk. Backstrom and Doos (1998) found that about 25% of injuries involving automated machinery were preceded by a disturbance in process material flow such as a piece of material becoming stuck, crooked or getting in an otherwise faulty position. In the US, lockout is required when an employee is required to place any part of his or her body into an area on a machine or piece of equipment where work is actually performed upon the material being processed (point of operation) or where an associated danger zone exists during a machine operating cycle. Minor tool changes and adjustments, and other minor servicing activities, which take place during normal production operations, are not covered by the OSHA lockout standard if they are routine, repetitive, and integral to the use of the equipment for production, *provided that the work is performed using alternative measures which provide effective protection* [emphasis added] (OSHA,1989). Machine control systems that can automatically detect and respond to hazardous operating conditions could be a way to reduce machine-related injury and fatality rates. Accurate laboratory testing criteria, based on key patterns manifested by the machine during hazardous events are vital to developing such new safety technology. This paper discusses safety design challenges in designing reliable safeguarding associated with machinery.

INTRODUCTION

Too many workers are killed on the job every year when their tasks bring them dangerously close to the hazards of running machinery (Pratt et al., 1996). Between 1992 and 1998, fatalities resulting from workers being caught in all types of machinery averaged 150 per year, with 189 fatalities in 1997 (Bureau of Labor Statistics (BLS), Census of Fatal Occupational Injuries (CFOI), 1992-1998). Also, between 1992-1996, serious nonfatal injuries resulting from workers being caught in machinery totaled 34,000 and were the leading cause of amputations among private industry wage and salary workers.

Human intervention into machinery systems often happens when material jams while it is being fed into the machine. Fully 10% of BLS CFOI narratives clearly specified that jammed machinery was involved and could easily have been a factor for other cases that did not have complete narratives. Although much is automated on modern machinery, most machines cannot unjam themselves. Workers must do it. Workers in the agriculture, manufacturing, and recycling industries (Table 1) are especially vulnerable (Windau, 1998).

Table 1 Fatalities in Some Machines Where Feedstock Often Jams (BLS, CFOI, 1992-1998)

Industry	Machine type that often jams	Total Fatalities
Recycling	Balers	31
Agriculture	Harvesting machinery	201
Manufacturing	Presses (injection molding, etc.)	49

A review of NIOSH's National Traumatic Occupational Fatality (NTOF) database (1980-1992) identified 27 fatalities involving stationary compactors and balers. Failure to lock out energy sources before clearing jams appear to be a factor in many of these incidents (NIOSH, 1999). NIOSH Fatality Assessment and Control Evaluation (FACE) investigations of more recent fatalities revealed two specific incidents in which personnel were inside the equipment and not following lockout procedures (NIOSH, FACE URL reference).

A machine is not necessarily safe when it is stopped (motionless). From a *machine operator's perspective*, a machine "stop" exists when a jam interrupts production flow. Too often, the focus of attention is solely on getting the system back into production, resulting in a restart of dangerous motion the instant the jam is cleared. From a *safety control system perspective*, a machine "stop" is achieved through the states of logic devices in the control system. Restart of a well-designed safety control can only happen when the logic devices return to the safe-operation state. A safety instrumented system (SIS) will read sensors, do calculations, and perform the logic required to identify potentially dangerous conditions, and generate outputs to actuators designed to mitigate the dangerous situation (Goble, 1998). Unfortunately, not all control system logic accounts for the undesired presence of a worker in the point of operation. Finally, from a *hazardous energy perspective*, "stop" includes locking out and relieving primary energy sources. Thereby, no energy is present and no injury is possible.

Descriptions of jams in balers

On large-scale recycling balers, several types of jams can stop production. An operator is expected to respond to these and return the baler to production.

> *Over-size bale: Too much material in the baling chamber.* On hand-tied bales, when too much material wedges into the baling chamber, wire cannot be wrapped around the bale. Someone must access the loading chamber to remove material so that a properly sized bale can be wrapped.
>
> *Over-size material loaded: Too much heavy material at the shear zone.* At the shear zone in a baler, the ram meets the open corner of the baling chamber. If there is too much material in this shear zone, the ram cannot shear through it as it

advances. Hydraulic system over-pressure relief causes the ram to reverse. Someone must access the loading chamber to remove this material by hand. Loading smaller amounts can avoid this.

Material bridges the feed chute. A person may try to jump on this bridged material from above to force it into the loading chamber. Ram motion can start automatically when sufficient material is in the loading chamber to block the light path of the photo-electric sensor. A person falling from above can interrupt the light path and start a cycle.

Potential causes of machine motion

While a person is clearing a jam, machine motion can occur for several reasons if power has not been adequately removed. Two types of energy need to be removed; electrical energy to pumps and directional control valves; and hydraulic energy stored in accumulators. Degraded operation of the following three components of a hydraulic circuit can have direct influence on the safety of personnel entering the point of operation (Foszcz, 1997).

1) Cylinder seal problems Hazardous motion can result should the cylinder ram drift (not hold in the position where it was initially stopped). Possible causes for this include: the directional control valve spool does not center; the valve spool does not shift completely; the valve spool or body are worn; the valve seats leak; or the piston seals leak. A cylinder may also drop a load being lifted even though the valve spool remains centered. This condition can result from: loose lines between the cylinder and valve; leaking or damage to the lockout or holding valve; broken springs on the lockout or holding valve; or leaking at the relief valve. Locking cylinders are a design option to consider.

2) Hydraulic direction-control valve problems Directional control valves direct or prevent flow through selected passages. Solenoids, hydraulic pilots, or mechanical linkages are used to shift the valve spool, which changes flow direction. Hazardous conditions arise when faulty or incomplete shifting occurs. Possible causes include: worn or binding control linkage; insufficient pilot pressure; burned out or faulty solenoid; defective centering spring; and improper valve spool adjustment.

3) Failure to discharge accumulator pressure Accumulators are pressure vessels in which hydraulic energy is stored and then released to even out the demand on the pump in circuits with uneven loading. The potential energy in an accumulator, when released into the system, can cause dangerous machine motion. In any accumulator circuit, a means should be available to automatically unload the accumulator when the machine is shut down. One way of accomplishing this is with a solenoid-operated 2-way valve that releases accumulator pressure to the reservoir when power is removed from the electrical control circuit. The solenoid must be energized to keep the valve closed.

RISK REDUCTION STRATEGIES

Given that the jam clearing task is high risk, highly reliable protective measures are appropriate. The risk control hierarchy is to 1) eliminate the hazardous task; i.e., no jams or fewer jams; 2) provide a well-engineered safeguard; i.e., not easily bypassed; and 3) provide worker training. While well-trained and supervised personnel can be a solution, worker protection should not rely on training alone as the primary injury prevention method.

The current solution to dangerous restart of machinery is lockout/tagout training and implementation (Grund, 1995). Lockout/tagout required by OSHA (Standard 29 CFR 1910.147), is the protective procedure workers must perform to remove hazardous energy from machinery prior to performing maintenance and servicing tasks. All energy sources for equipment must be turned off or disconnected and the disconnecting device, switch, or valve locked and labeled with a tag warning not to energize the machinery. However, in 1998, lockout/tagout violations ranked fourth on OSHA's top 10 list of most frequently cited standards (3,532 violations) and fourth on the top 10 list of serious violations (2,537 serious violations). Also, minor tool changes and adjustments, and other minor servicing activities, which take place during normal production operations, are not covered by the OSHA lockout standard if they are routine, repetitive, and integral to the use of the equipment for production, *provided that the work is performed using alternative measures which provide effective protection* [emphasis added].

Because of these shortcomings with the training solution, the two strategies that are higher on the risk control hierarchy warrant further consideration. NIOSH and industry partners, within the framework of the National Occupational Research Agenda (NORA), have developed priorities for finding such alternative measures (NIOSH, 1998). The NORA Traumatic Injury research agenda calls for:

> 1) designing workplaces, machines, and equipment to remove the possibility of making unsafe choices or actions (engineer-out hazards); and
>
> 2) in designing and modifying workplaces to improve safety, "passive controls," those which do not require choices or actions by workers in order to be protective, should take priority over "active controls," which require workers to take action or change behavior.

ELIMINATING DANGEROUS PRODUCTION DISTURBANCES

Making machines less dangerous often is viewed as a function of new machinery design. But the metalforming industry, as an example, is adding new monitoring capabilities to existing presses to eliminate disturbances due to part feeding problems. Press load monitors have been used for more than 30 years to protect stamping presses and dies from damage due to overloads (Kuvin, 2001). These devices can also effectively reduce operator exposure to jam-clearing hazards. A normal load signature is established by the press manufacturer. During operation, the device compares the normal load signature with the actual force at that point in the stroke. Should the force exceed the normal value anywhere in the stroke, instantaneous machine shutdown is initiated to stop the press before the machine is damaged or a dangerous condition occurs. A great deal of benchtesting is needed to establish a safe-signature threshold and appropriate monitor locations. At its most sophisticated level, data can be fed back to the process controller to make automatic adjustments to compensate for the out-of-tolerance conditions.

In agriculture, jams form in somewhat different ways, but present similar hazards during jam-clearing. The buildup of jams in the feed rolls of harvesting machinery such as hay balers, combines, and corn pickers depends on the forward speed at which the implement is gathering in crop stalks and the intake of extraneous weeds. Reversers on feed mechanisms on combines provide workers with a means to clear jams from the safety of the operator's control station. Interlocks are also required, where feasible.

FUTURE DIRECTIONS

Two directions in design improvement to mitigate the hazards of the jam-clearing task are indicated: 1) continue to install safeguarding that is not prone to human error/ misuse; and 2) develop better process monitors and controls that design out, or greatly reduce, the hazardous task exposure. Injury prevention efforts should be directed toward putting countermeasures in these categories into regular practice.

with lockout in a manner that avoids human error. In a trapped key control system, power must be shut off before access can be gained through an interlocked door. Both the door and the power switch incorporate a lock. A removable key is trapped in either the door lock or the switch lock. The lock on the door is arranged so that the key can only be released when the door has been closed and locked. This allows transfer of the key from the door to the switch lock. Closing the switch traps the key, so that it cannot be removed while the switch is in the on position. A key exchange box is used if several sources of power are involved. However, to be most effective, duplicate keys should not be permitted.

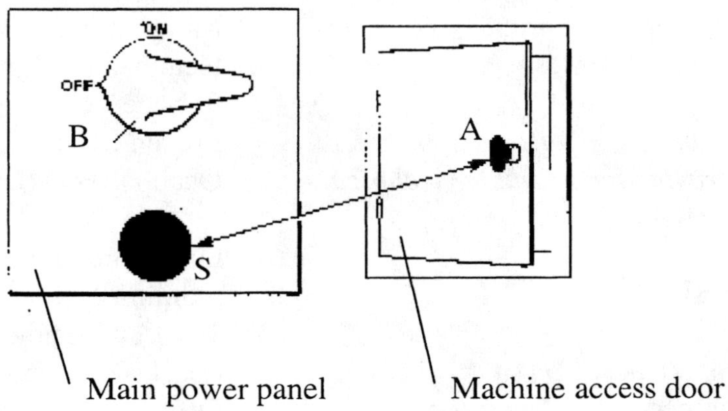

Figure 1 Key A is trapped in the door and can only be turned and removed when the door is closed, thus locking the door. Switch B cannot be turned to the ON position until key A has been inserted, turned, and trapped in lock S. Key A can only be removed from S when B is in the OFF position.

Safeguarding that removes the possibility of making unsafe choices

The trapped key method (Figure 1) is designed to combine the safety control stop

Hazard elimination or reduction

Designing machine operation monitors that stop material processing and in-feed before a jam can form are the highest-level safety solution. An easier-to-clear condition that can be accomplished from outside the danger zone eliminates the high-level hazard and replaces it with a much reduced one. Importantly, exposure time is reduced, or even eliminated. If some exposure, for a shorter period of time is necessary, then the monitor may be designed to automatically initiate the lockout process via the emergency-stop circuit of the machine.

Two design directions for making the jam-clearing task safer have been advanced in the foregoing discussion. The degree to which these design challenges are met, and design interventions put into practice, will have a direct bearing on the avoidance of future machine-related injuries and fatalities in the recycling, manufacturing, and agriculture industries. For its part, NIOSH will continue to follow the NORA research priorities that require laboratory testing and evaluation of new safety devices intended to protect against typical normal operation, caught-in-machinery hazards.

Acknowledgement: Darlene Weaver, PTB/DSR, provided valuable technical support in this research.

REFERENCES

Backstrom, T and Doos, M [1998] "Moving Parts of Machinery", in Encyclopedia of Occupational Safety and Health, ed. J. Stellman, International Labor Office, Geneva, 58.1-58.11

BLS, Census of Fatal Occupational Injuries. Washington, D.C.: U.S. Department of Labor, Bureau of Labor Statistics. 1992-1998. http://stats.bls.gov/oshhome.htm

Foszcz, J [1997] Basics of Actuating Cylinders, *Plant Engineering*, July.

Goble, W [1998] Control Systems Safety Evaluation and Reliability, Instrument Society of America, Research Triangle Park, NC.

Grund, E [1995] Lockout/Tagout: The Process of Controlling Hazardous Energy, NSC Press, Itasca, IL.

Kuvin, B [2001], "Sensors and Signatures Guide Press Control", *Metal Forming*, May, pp 24-34.

NIOSH, FACE Reports
http://www.cdc.gov/niosh/face/faceweb.html

NIOSH [1998] Traumatic Occupational Research Needs and Priorities, DHHS (NIOSH) PUBLICATION No. 98-134, Cincinnati, OH.

NIOSH, NTOF [1999]. National Traumatic Occupational Fatalities Surveillance System. Morgantown, WV: U.S. Department of Health and Human Services, Public Health Service, Centers for Disease Control and Prevention, National Institute for Occupational Safety and Health. Database.

Occupational Safety and Health Administration [1989] OSHA 29 CFR 1910.147, The Control of Hazardous Energy (Lockout/Tagout), Government Printing Office, Washington, DC.

Pratt, S, Kisner, S, and Helmcamp, J [1996] "Machinery-related Occupational Fatalities in the United States, 1980 to 1989". *Journal of Occupational and Environmental Medicine*, 38(1), pp 70-76.

Windau, JA [1998] "Worker Fatalities from being Caught in Machinery", *Compensation and Working Conditions*, Bureau of Labor Statistics, Winter, pp. 35-38.

Proceedings of
2001 ASME International Mechanical Engineering Congress and Exposition
November 11–16, 2001, New York, NY
SERA-Vol. 11

IMECE2001/SERA-24005

International Safety Alert Symbol

Ralph L. Barnett
Professor, Mechanical And Aerospace Engineering
Illinois Institute of Technology
Chicago, Illinois

Chairman
Triodyne Inc.
Niles, Illinois

Raymond Wambaja
Mechanical Engineer
Triodyne Inc.
Niles, Illinois

ABSTRACT

With the adoption of the international safety alert symbol, the safety profession has lost an important weapon in the war against injury. The symbol is not uniquely associated with safety, it does not have an optimum shape and it has no intrinsic pictorial to communicate danger to untrained people from every culture. The symbol represents a tragic "missed opportunity" for mobilizing personal vigilance.

I. INTRODUCTION

In 1968, Margaret Mead and Rudolph Modley [1] noted that people spoke at least twenty-eight hundred languages. These people are exposed to all of the dangers associated with technology and nature. Because the unwary and untrained are always in jeopardy, motivation exists for the development of effective safety communication systems. This, in turn, gives rise to the establishment of a universally understood safety alert symbol [2].

Symbols provide a number of advantages to safety professionals, e.g.,

- A symbol engenders a faster reaction than any combination of words and/or symbols [3].
- Symbols are language independent and require less space than text [4].
- On average, the threshold size for symbolic signs was approximately half that of the corresponding text version [5].
- A symbol sign can withstand greater degradation and still be recognizable [6].
- Warnings printed conspicuously and paired with pictorial icons would produce greater comprehension and memory than would warnings in plain print without pictorial icons [7].
- A symbol used as an icon attracts attention to warnings and safety related matters [8].

Safety related issues are sometimes overt and sometimes subtle. A safety alert symbol serves to flag safety problems and activate personal vigilance. For this reason, the safety alert symbol (Fig. 1) was developed and standardized.

II. BACKGROUND

The earliest written languages of Egypt, China and Mesopotamia used pictures to represent ideas. It seems ironic that today we are returning to our roots after developing a modern alphabet that has evolved far beyond the simple representation of familiar objects [9].

Figure 1 - International Safety Alert Symbol

A. Highway Signs

In the twentieth century the most prevalent use of symbols has been for highway signs [10]. One of the first attempts at standardizing symbols for roadways was undertaken in Paris in 1909 at the Convention with Respect to the International Circulation of Motor Vehicles. Standard symbols were developed to eliminate language barriers and communicate messages universally and simply [11]. The triangle was approved as the shape for warning signs in 1926 at another convention on motor traffic. The triangular warning sign was standardized by the United Nations in 1949 in their protocol on Road Signs and Symbols. The symbol is used throughout Europe.

The original use of the arbitrary exclamation point for warning was in the highway context. (Ref. 2)

B. Agricultural, Construction and Industrial Equipment

The tractor technical committee of the Society of Automotive Engineers (SAE) approved the safety alert symbol shown in Fig. 1 in December 1972. It was introduced in the SAE Recommended Practice: SAE J284a - Safety Alert Symbol for Agricultural, Construction and Industrial Equipment. The general purpose nature of this recommended practice may be inferred from the concise language in its purpose, scope, description and applications, to wit:

"1. Purpose

1.1 To establish a safety alert symbol for use on agricultural, construction and industrial equipment.

1.2 To provide a symbol which means ATTENTION! BECOME ALERT! YOUR SAFETY IS INVOLVED!

2. Scope - This recommended practice presents the general uses, limitations on use, and appearance of the safety alert symbol.

3. Description

3.1 The safety alert symbol shall be an equilateral triangle with rounded corners and with an "exclamation mark" located in the center as shown in Fig. 1. The dimensions are optional.

3.2 The safety alert symbol should be of contrasting colors which will cause the "exclamation mark" to stand out.

4. Applications

4.1 The symbol should be used as:

4.1.1 A symbol which can be used in conjunction with warning statements and signs.

4.1.2 A symbol in instruction manuals.

4.1.3 A symbol in connection with agricultural, construction and industrial equipment safety standards.

4.1.4 A symbol to appear on communications which concern agricultural, construction and industrial equipment safety.

4.2 The symbol should not be used:

4.2.1 To indicate safety compliance or a safety characteristic.

4.2.2 Alone on equipment for safety purposes."

It should be noted that the American Society of Agricultural Engineers (ASAE) adopted SAE J284 in 1972. It is currently part of the ASAE standards.

C. American National Standards Institute (ANSI)

On July 24, 1980, a draft of ANSI Z535.4, Product Safety Signs and Labels, adopts the SAE J284a Safety Alert Symbol in Fig. 1. In 1991 the symbol was specified in both ANSI Z535.3, Criteria for Safety Symbols, and Z535.4-1991. Its purpose is to indicate a potential personal safety hazard [12].

D. Symbol Sourcebook

In 1972, Henry Dreyfuss published an authoritative guide to international graphic symbols [2]. The safety alert symbol in Fig. 1 is shown under *Safety* and under *Graphic Form*. Further, the triangle is featured without the exclamation point in a number of safety related applications.

E. Handbook of Pictorial Symbols

Rudolph Modley, in 1976, published a handbook containing 3,250 examples of symbols from international sources [13]. The safety alert symbol, ⚠, is presented as a Danger symbol (Ibid. pp. 110, 132). More specific safety signs are developed by replacing the exclamation point with a dedicated pictogram. Examples of triangular safety symbols are illustrated throughout Modley's handbook (Ibid. pp. 72, 108, 132).

III. SIGN GEOMETRY

Safety signs are typically fashioned by the artistically challenged. Their world does not extend beyond the straightedge and the compass. This accounts for the predominant use of regular polyhedra such as the equilateral triangle, the square, the hexagon and the octagon used for roadway stop signs. The rectangle is very popular; the trapezoid is not. Circles have been adopted for railway demarcation and, when used in conjunction with a single diagonal slash, connote a prohibition against any pictogram included within the circular border. All safety signs have a convex shape and are simply connected (mathematical term).

Safety signs may be contrasted with advertising signs and product labels, whose creators are highly skilled graphic designers who use free form curves and combinations of convex and concave borders (e.g. the valentine heart). They may use holes and other forms of multiply connected labels. Stick figures are seldom used; colored likenesses are.

Riley, Cochran and Ballard, in 1982, investigated 19 simple geometric shapes of warning labels to establish the preferred shapes for warning indicators [14]. Table 1 defines the shapes and presents them in decreasing order of preference. The ranking shows that when shape alone is the determining factor in describing a warning, an equilateral triangle on its vertex is the preferred warning shape. This means that the widely used safety alert symbol in Fig. 1 is upside-down! The study suggests that shapes that appear unstable tend to be preferred as warnings.

Another observation from Table 1 seems appropriate. The worst candidate, 19 out of 19, is the six sided tag shaped figure; this is the most popular shape used for lockout/tagout applications [15].

Table 1 - Preferred Shapes for Warning Indicators

Ranking	Shape	Description
1	▽	Equilateral triangle pointing downward (traffic yield shape)
2.5	◇	Square on a point or diamond (traffic warning sign)
2.5		Octagon (traffic stop sign shape)
4.5	▷	Equilateral triangle pointing right (traffic no passing shape)
4.5	△	Equilateral triangle pointing upward (traffic yield shape)
6	⬡	Hexagon
7		Six-sided figure (slow moving vehicle sign shape)
8.5	◁	Equilateral triangle pointing left
8.5	⬨	Parallelogram on a point
10	⬠	Pentagon (similar to school crossing roadway sign)
11.5	⏢	Trapezoid on its long base
11.5		Trapezoid on its short base
13	⬭	Parallelogram on its side
14	○	Circle (railroad crossing roadway sign shape)
15	▭	Rectangle on its long base (similar to traffic information sign)
16.5	□	Square on its base
16.5	⬭	Ellipse on its major axis
18	▯	Rectangle on its short base (traffic regulator sign)
19		Six-sided figure (tag shape)

IV. UNIQUENESS

The power of the safety alert symbol to highlight a safety concern is diluted when the symbol is used for a myriad of lesser tasks. Each non-safety appearance of the symbol produces an "anti-teaching" effect. It is, unfortunately, too late to restrict the symbol to safety related matters; its other uses are too widespread. For example:

1. The Fine Arts Library, Cornell University, uses the symbol for reserve materials:

> Fines for reserve material are steep. Please be sure to return your material promptly!

2. The Top Tools Company uses the symbol in their Requirements Check:

> Software:
> *You must have administrator privileges for the system where HP TopTools will be installed.*

3. The web site for Worldwide Riche$ (1998) uses the symbol to arrest the reader's attention:

> Successful web sites motivate potential clients *"to act"*. Expertise in Internet marketing & web site design are essential to your business's success.

4. The use of the symbol as a non-safety related icon may be found at http://www.gemini.edu/sciops/OThelp/bugs.html :

> ***Bugs and Limitations***

5. The Nicholas School of the Environment adopts the symbol in their course syllabus, NSOE ENV351 Course Outline:

> **Late lab assignments and projects will be significantly penalized!** Contact the instructor prior to the due date if you will not able to complete the assignments on time.

6. From a personal homepage for Dr. Kim Ainsworth-Darnell,

> **What is the world coming to? Even my Mom has a homepage! Check it out :)**

7. The symbol is used by the Boston Public Schools Office of Technology to announce newsworthy items:

> News and Happenings

8. The symbol is sometimes used in dictionaries to demarcate "words that are likely to cause embarrassment or anger if they are used in the wrong situation." The following comment is from Melody IS., 1997 -1999 HME Media [16].

> **24:38** · Both the **exclamation mark** and the **triangle** are often used for warnings. The combination of them is definately a sign *warning* for a danger. ⚠ is a traffic warning sign, but this ideogram is also used, for instance, in Oxford dictionaries to warn for *taboo words.*

V. REPRESENTATIONAL SYMBOLS

There are three categories of symbols that have been identified by both Modley [13] and Dreyfuss [2]: representational or image related, abstract or concept related, and arbitrary. Examples of these are illustrated in Fig. 2. Clearly, the safety alert symbol falls into the arbitrary category.

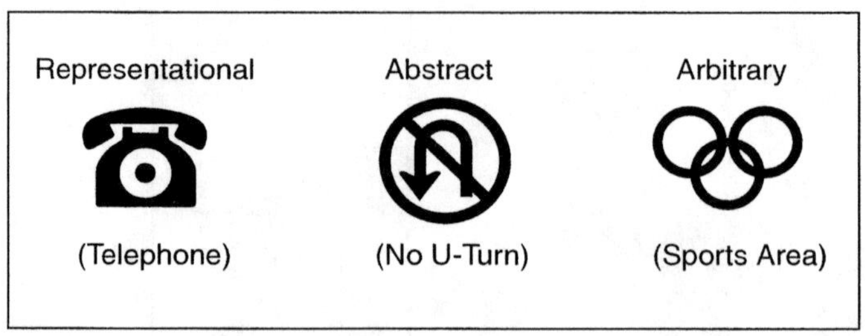

"Arbitrary symbols are those where the meaning is arbitrarily assigned to the graphic. Thus an exclamation mark does not depict any aspect of the real world object, but has a meaning that has been arbitrarily assigned to it and that we have had to learn to understand without any assistance from representational elements within the graphic itself. Where such arbitrary symbols are used the graphic image should be

simple and clearly distinct from any other symbol that might be used in a similar context." Edgeworthy [6]

In summary, the safety alert symbol has no intrinsic meaning; every viewer must be trained to take advantage of its safety significance. Furthermore, it is neither distinct nor unique.

For the sake of concreteness, the authors have prepared the safety alert symbol in Fig. 3 in various sizes. The triangle with its downward facing vertex was selected since it is the optimum shape; furthermore, the particular collection of graphics is clearly unique. An attempt was made to depict a general image of danger as opposed to a specific accident scenario. The crossed crutches are supposed to imply a leg injury; an eye injury is represented; a head fracture is illustrated; and finally, a death skull with a pained expression is included. It was hoped that this combination of maladies would not suggest a common cause and that the smorgasbord of mischief would imply a general notion of danger. Whether this goal was achieved with this symbol is not important. What is important is that some symbol attain such a goal so that all viewers will perceive a danger communication.

With respect to the proposed safety alert symbol shown in Fig. 3, there are some characteristics that should be noted:

- The salient features of the graphic symbol are preserved in the smaller sizes.
- The symbol retains some classic characteristics of the skull and crossbones used in the old poison symbol.
- Both injury and death are represented in the graphic.
- Because the symbol has not been tested, it is not known whether its imagery will translate among cultures.
- Either danger or its antonym, safety, could have been used as a safety alert symbol.

The development of an effective general danger graphic is a sophisticated task that should be undertaken by the safety profession. If you imagine a protocol that assembles focus groups from every culture, this will give some indication of the magnitude of such a project. In the meantime a little doodling seems in order.

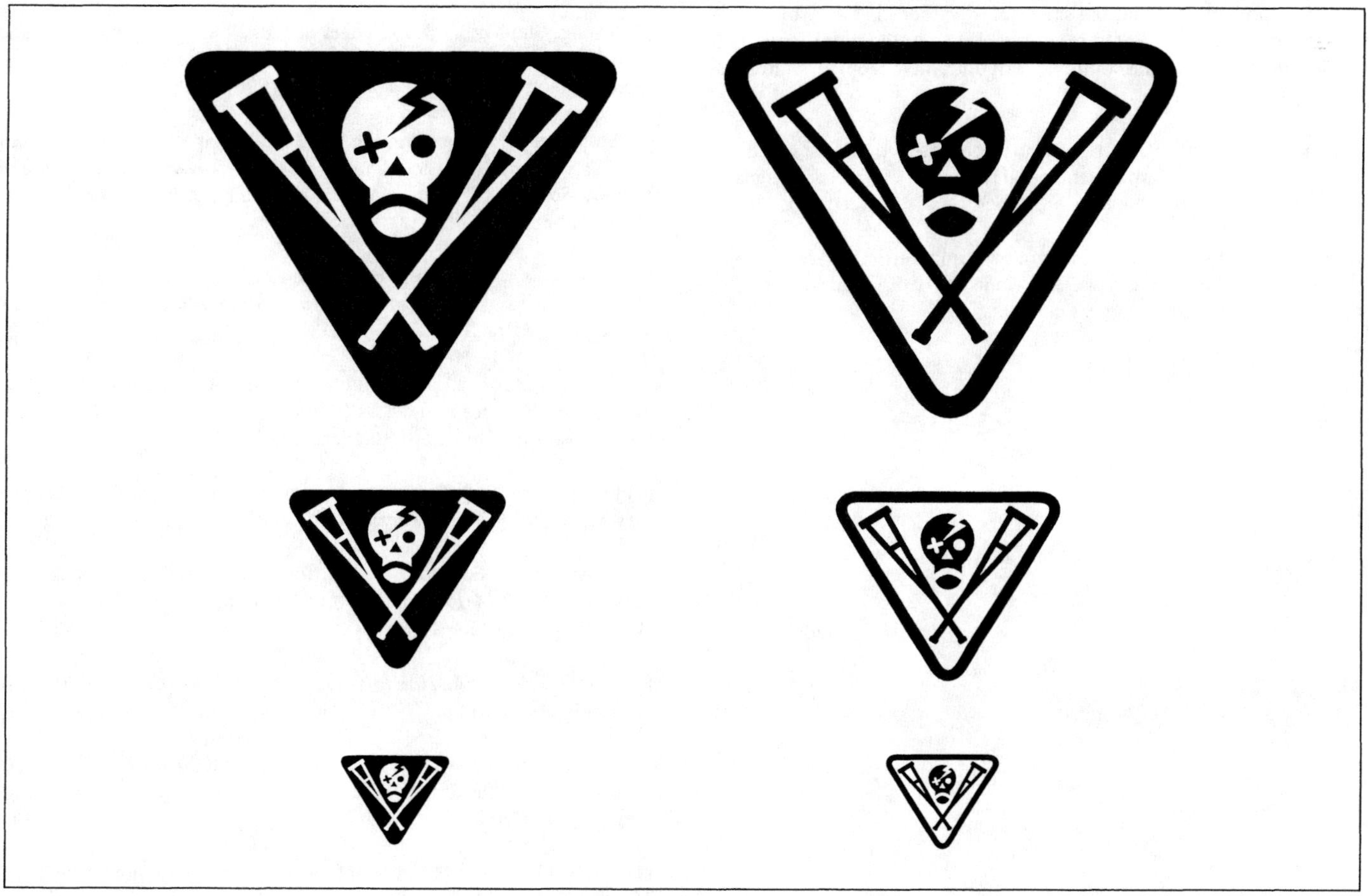

Figure 3 - Proposed International Safety Alert Symbol

VI. CONCLUSIONS

1. The international safety alert symbol depicted in Fig. 1 evolved through the process of consensus; it is not the product of research.

2. Research shows that the shape of the safety alert symbol is not optimal; indeed, it is upside-down. We have standardized an inferior shape.

3. As a further object lesson in research vs. consensus, we note that the least desirable warning shape has been reserved for the safety lockout tag.

4. The safety alert symbol enjoys widespread use in dozens of non-safety related applications. This may be explained, in part, by the fact that no talent is required to draw the symbol.

5. Because the safety alert symbol is an "arbitrary" graphic, its meaning must be taught. This teaching effort is diluted by the proliferation of uses outside of the safety arena.

6. It is reprehensible that the safety alert symbol is an "arbitrary" symbol without intrinsic meaning. It cannot communicate the notion of safety to the community of untrained people that dominate the world.

7. An untested safety alert symbol is proposed that has an optimum shape and a unique image that attempts to reflect a sense of danger without focusing on a specific safety scenario.

8. To restrict the application of any new symbol to safety-related issues, it will be necessary for one of the safety organizations to appeal to intellectual property attorneys.

REFERENCES

[1] Mead, Margaret and Rudolph Modley, "Communications Among All People, Everywhere," *Natural History*, V. 77 #7 (1968): 56-63.

[2] Dreyfuss, Henry, *Symbol Sourcebook,* New York: McGraw Hill, 1972. pp. 140, 143, 186-187.

[3] Janda, H.F. and Wayne N. Volk, "Effectiveness of Various Highway Signs," *Proceedings of the Fourth Annual Meeting of the Highway Research Board*, Washington, D.C., December 1934.

[4] Zevega, Harm J., "Comprehensibility Estimates of Public Information Symbols; Their Validity and Use," *Proceedings of the Human Factors Society 33rd Annual Meeting*, Denver, October 16-20, 1989, p. 979.

[5] Kline, Donald W. and Perry Fuchs, "The Visibility of Symbolic Highway Signs Can Be Increased Among Drivers of All Ages," *Human Factors* v. 35 (1993): 29.

[6] Edgworthy, Judy and Austin Adams, *Warning Design: A Research Prospective*, Bristol, PA: Taylor & Francis, 1996, pp. 76-77.

[7] Young, Stephen L. and Michael S. Wogalter, "Comprehension and Memory of Instruction Manual and Warnings: Conspicuous Print and Pictorial Icons," *Human Factors* v. 32 (1990): 639.

[8] Young, S.L., "Increasing the Noticeability of Warnings: Effects of Pictorial, Color, Signal Icon, and Border," *Proceedings of the Human Factors Society 35th Annual Meeting*, San Francisco, September 2-6, 1991, pp. 580-584.

[9] Giedion, G., "Symbolic Expression in Prehistory and in the First High Civilizations," in *Sign, Image and Symbol*, G. Kepes (ed.) New York: George Braziller, 1996, pp. 78-91.

[10] Collins, Belinda Lowenhaupt, "The Development and Evaluation of Effective Symbol Signs," *NBS Building Science Series 141*, Washington: National Bureau of Standards, 1982, p. 5.

[11] King, L.E., "A Laboratory Comparison of Symbol and Word Roadway Signs," *Traffic Engineering and Control* v. 12 (1971): 518-520.

[12] "American National Standard for Product Safety Signs and Labels," *ANSI Z535.4-1991,* New York: American National Standards Institute, approved June 1991, pp. 2-3.

[13] Modley, Rudolph, *Handbook of Pictorial Symbols*, New York: Dover Publications, 1976.

[14] Riley, Michael W., David J. Cochran and John L. Ballard, "An Investigation of Preferred Shapes for Warning Labels," *Human Factors* v. 24 #6 (1982): 737-742.

[15] "The Control of Hazardous Energy (Lockout/Tagout)," 29CFR 1910.147, Occupational Safety and Health Administration, Washington, D.C., July, 1994.

[16] http://www.symbol.com/encyclopedia/242438.html

Proceedings of
2001 ASME International Mechanical Engineering Congress and Exposition
November 11–16, 2001, New York, NY
SERA-Vol. 11

IMECE2001/SERA-24006

THE RISK BASED ALLOCATION OF SAFETY INTEGRITY LEVELS FOR MACHINERY CONTROL SYSTEMS & THE INTERNATIONAL STANDARD IEC 62061

Nicola Worsell, Health & Safety Laboratory
Broad Lane, Sheffield, S3 7HQ, UK,
nicola.worsell@hsl.gov.uk
Tel: +44 (0) 114 289 2572 Fax: 114 289 2444

Jill Wilday, Health & Safety Laboratory
Harpur Hill, Buxton, UK
jill.wilday@hsl.gov.uk
Tel: +44 (0) 114 289 2156 Fax: 114 289 2160

Colin Chambers, Health & Safety Laboratory
Broad lane, Sheffield, S3 7HQ, UK,
colin.chambers@hsl.gov.uk
Tel: +44 (0) 114 289 2568 Fax: 114 289 2444

ABSTRACT

This paper outlines the draft international standard IEC 62061 – Safety of machinery – Functional safety – Electrical, electronic and programmable electronic control systems [1]; this is intended to be a machinery sector implementation of the basic safety publication IEC 61508 – Functional safety of electrical/electronic/ programmable electronic safety-related systems [2]. A fundamental concept of IEC 61508 is the definition of safety performance criteria for safety functions implemented by safety-related electrical, electronic and programmable electronic systems in terms of safety integrity levels (SILs). This paper, therefore, goes on to describe a methodology, currently under development, for the risk based selection of SILs that it is proposed should be incorporated into an informative annex of IEC 62061.

INTRODUCTION

The International Electrotechnical Commission standard IEC 62061 will be a machinery sector implementation of the basic safety publication IEC 61508 – Functional safety of electrical/ electronic/programmable electronic safety-related systems [2]. It is intended to enable designers, manufacturers, suppliers and users to develop and validate safety-related electrical, electronic and programmable electronic (E/E/PE) control systems that adequately reflect their contribution to risk reduction at a machine.

IEC 61508 is generic across all industries. A fundamental concept of this standard is the definition of safety performance criteria for safety functions implemented by safety-related electrical, electronic and programmable electronic systems in terms of safety integrity levels (SILs). Because of this standard's status as a basic safety publication it is necessary for the machinery sector standard IEC 62061 to also specify safety performance in terms of SILs.

Whilst the principles of risk assessment are currently set out in ISO 14121 (EN 1050) they do not contain guidance on the practical application of risk assessment to machinery. Furthermore, for a machinery designer to comply with IEC 61508 they first need to be able to identify the required safety-related control functions of a machinery control system and then allocate the appropriate SILs to these functions. This SIL is a target failure measure for the safety-related E/E/PE control system.

SILs are, in part, described in terms of decade ranges of probability of failure to danger. Unfortunately, there is currently no standardized methodology in the machinery sector for risk estimation expressed in terms of probability of harm (failure to danger). The working group mandated to produce IEC 62061, have therefore been obliged to develop a risk based methodology for the allocation of SILs. It follows similar principles and techniques for probability estimation to those used in other sectors, for example the chemical process [12].

A project team at the Health and Safety Laboratory (HSL), an in-house research agency of the UK Health and Safety Executive (HSE) were contracted by HSE on a project to develop a risk assessment methodology for use during machinery design for several years. The project team provided information on this methodology to the working group who were given the task of developing an informative annex on the risk based selection of safety integrity levels for IEC 62061. At that time, the standard was in the very early stages of development. A Committee Draft (CD) for Consultation was published in September 2000; the deadline for comments was

January 2001. Since then this methodology has been extensively revised to make it more specific to the needs of IEC 62061. A further CD is expected to be published by the end of 2001. The methodology described in this paper is therefore almost certainly likely to have been further revised by the time this paper is presented. In order to ensure that the reader has the most up to date version of the methodology they are advised to obtain a copy of the CD when it is published.

OVERVIEW OF DRAFT IEC 62061

Whilst some sections of the standard are written in detail others are written more in the form of an outline, with the detail yet to be agreed. All clauses are currently being revised and clauses 4, 5 and 6 extensively expanded. The Annexes are as yet incomplete. Whilst the framework and underlying ideas as described below are likely to remain the same much of the detail is expected to have changed significantly when the next CD is published towards the end of 2001. The draft standard as published in the September 2001 CD contains the following clauses.

1. Scope

The scope of the standard is limited to a consideration of the safety-related control functions implemented by an electrical/electronic or programmable electronic control system that reduces the risk of injury or damage to the health of persons in the immediate vicinity of the machine and those directly involved in the use of the machine. It is restricted to those risks arising directly from the hazards of the machine itself or from a group of machines working together in a coordinated manner. It does not specify requirements for the performance of non electrical elements although it allows for the amount of risk reduction these elements provide to be taken into consideration. Finally it does not cover electrical hazards arising from the control equipment itself such as electric shock.

2. Normative references

Other than relevant parts of IEC 61508 a number of machinery sector standards are given as normative references. One of the major challenges facing the team developing IEC 62061 will be in ensuring that the requirements of IEC 61508 are met, whilst at the same time, not straying from the concepts contained within machinery sector standards, in particular:

ISO 12100 - Basic concepts, general principles for design;

ISO 14121 - Principles for risk assessment;

EN 954 - Safety related parts of control systems, General principles for design.

3. Definitions & Abbreviations

An extensive list of definitions are included in the standard, that for consistency within the sector, draws not only upon those given in IEC 61508 but also ISO 12100.

4. Management of functional safety

This clause outlines the responsibilities of management for the provision of suitable resources, recording and maintenance of information, for example, verification and validation plans.

5. Determination of functional safety

This is where the functional safety requirements of the machinery control system are determined by drawing together the principles of good design and the use of risk assessment. There are many situations on a machine where electrical control systems of machines have been provided to achieve risk reduction. One of the most typical is the use of an interlocking guard where, when it is opened to allow access to machine parts that are hazardous by virtue of their movement, it signals the electrical control system to inhibit hazardous machine operation. This clause, if only briefly at present, covers issues such as the need to provide information about the machine, its use, foreseeable misuse, (including defeating of safeguards) and machine limits. It outlines the requirements for hazard identification, risk estimation and evaluation and a strategy for risk reduction but references an informative annex on risk assessment for the detail on how to carry out risk assessment in practice. This clause also goes into some detail about the stipulation for and contents of a control system safety requirement specification.

6. Design and integration

This clause describes how to select or design and implement the safety-related electrical control system in order to meet the functional and safety integrity requirements specified in the safety requirements specification described in clause 5. It deals with issues such as fault detection, behavior and system architecture.

7. Information for use

This describes the requirements for the provision of information for use, in accordance with the normative references.

8. Validation

This outlines the requirement for the validation of the safety-related electrical control system in accordance with the prepared plan referred to in clause 4.

Annexes

There are currently two annexes, both of which are informative not normative. Annex A refers to risk assessment and a methodology for the risk-based selection of SILs for safety-related control functions identified in clause 5. Annex B will contain worked examples. The rest of this paper describes the current version of a methodology for the risk based selection of SILs for defined safety-related control functions of E/E/PE control systems which it is proposed will be incorporated into Annex A.

RISK ACCORDING TO ISO 14121

The concept of SILs is new to the machinery sector which within Europe is still struggling to get to grips with the requirements of the machinery directive [1]. However, risk is defined for the machinery sector in ISO 14121 as a function of the severity of harm for the considered hazard and the probability of occurrence of that harm. Furthermore, the probability of occurrence of that harm is described as being made up of: the frequency and duration of exposure; the probability of the hazardous event; and the possibility to avoid or limit the harm. This is illustrated diagrammatically within ISO 14121 as shown in figure 1 below.

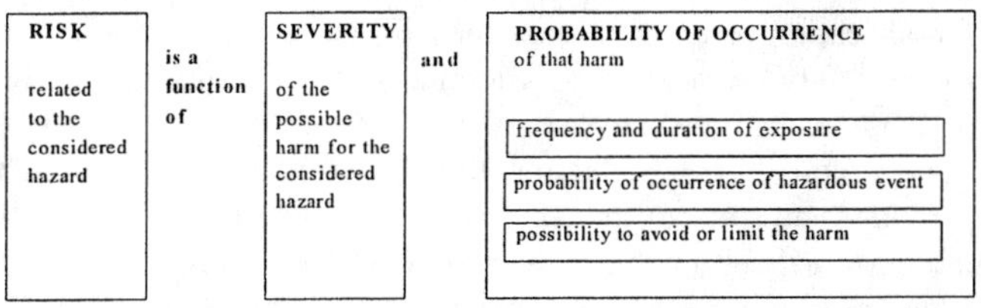

Figure 1: Risk according to ISO 14121

For machinery the severity of the possible harm (or consequences) from a specific hazard will, in the worst case, rarely lead to more than one fatality. However, in most cases the consequences will be injuries ranging from the severe and permanent, such as amputation, or serious, such as broken bones, to the less serious, such as, cuts and abrasions. This is unlike other industries, for instance, the chemical process or nuclear, at which IEC 61508 is also aimed, where the more serious accidents can involve large numbers of casualties. Moreover, one specific machinery hazardous event can be associated with a range of potential consequences. For example, entanglement and being drawn into a machine, as known to occur in the agricultural sector with trailed powered equipment and machine parts capable of seeing large torques powered via power tack off drives from the tractor engine and currently with no means of emergency stop, would most likely lead to a fatality but there is some chance that the victim of such an incident may survive, albeit with severe and permanent injuries. But, the chance of getting away with only minor injuries would be diminishingly small. An example at the other end of the consequence scale would perhaps be the ejection of a grinding debris or hot chips from a milling operation where a fatality is unknown but minor injuries relatively likely. For other machines there will be one injury in particular that will be the most likely. Nonetheless, there may be a small probability of more or less severe injuries. For example the failure of a light guard on a power press.

The frequency of exposure to a machinery hazard will be at least once a year and tends to be much higher, more often than not daily. Therefore the low demand situation described in IEC 61508 is not considered by IEC 62061. The duration of exposure can vary from a few seconds to continuous.

The probability of occurrence of the hazardous event will be directly related to the SIL of any safety-related control functions that act to prevent it. However, there may also be other technology measures that act in addition to, or association with the safety-related electrical control system. These therefore need to be taken into account, when allocating the appropriate SIL.

Finally, the possibility to avoid or limit the harm once the hazardous event has occurred needs to be taken into account. This can range from being impossible to there being a very good chance of avoiding or limiting the harm if an operator is given both warning and time to take action before the hazardous event causes harm.

DEVELOPMENT OF MODEL TO QUANTIFY RISK

In order to quantify this risk a fault tree model was developed. The probability of an accident of a particular severity occurring is a combination of the probability of the accident occurring and that it results in an injury of a particular severity level. In doing this account is taken of the possibility to avoid or limit the harm. This is represented by the fault tree shown in figure 2 below.

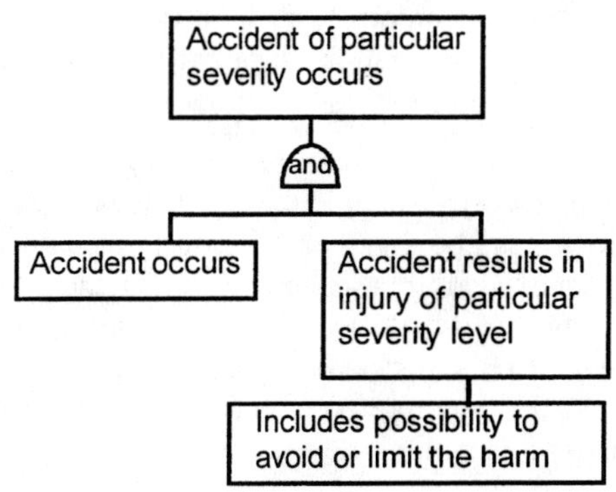

Figure 2: Probability of accident of particular severity

The accident will occur if the safety-related control function and any other technology designed to protect against it fails and if a person, not necessarily the operator, is situated in such a position that they can be harmed as a result of the hazardous event occurring. Any location where a person can be harmed is referred to as the danger zone. For a particular accident to occur the machine may also need to be operating in a particular mode. There is also a possibility that any faults present are revealed by a functional test (and corrected) before the hazardous event can occur. These considerations are represented by the fault tree shown in figure 3 below.

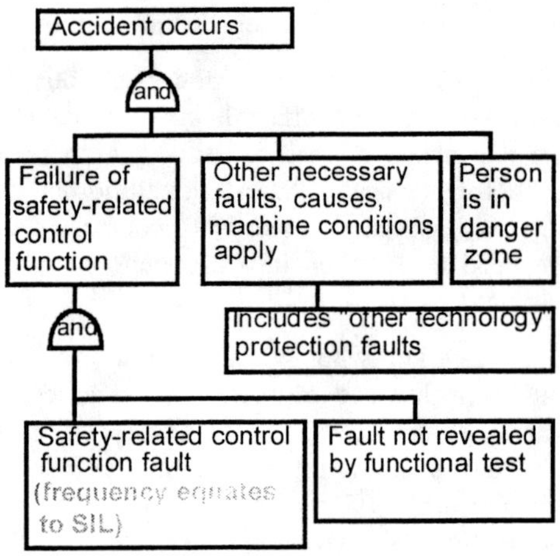

Figure 3: Probability that accident occurs

RISK BASED METHODOLOGY FOR SIL ALLOCATION

A methodology, designed to be used by a team, that assumes users may not be familiar with fault trees, is in the process of being developed. The user is required to collect certain information, and a series of forms are used to structure and document this information. Detailed step-by-step guidance on how to go about collecting the information is also being produced. Guidance along with look-up tables is provided to help the user with probability data and the arithmetic required is indicated on the forms.

The current version of this methodology is illustrated, by example, below. An example of what a completed set of forms may look like can be found on subsequent pages. The reader is, however, once again reminded that this methodology is under development and is subject to change. Furthermore, that the guidance given in this paper, whilst sufficient to adequately describe the methodology, is not comprehensive enough for the reader to use it in full.

The team is first required to describe the accident scenario and record this on the first form in accordance with the guidance given. For example, when a metal working power press is hand-fed, one of the main hazards to be eliminated by the design of safeguarding is that of any part of the operator's body being crushed between the tools. Without suitable safeguards this may happen, for example, whilst the operator is aligning the work piece. The safety-related control function is therefore to prevent the operation of the press whilst the operator's hands are in the danger zone. It is not necessary at this point to specify the nature of this function. This information is recorded on form 1 as shown overleaf.

The team then needs to brainstorm a list of all the underlying causes for the accident scenario under consideration. These are events or states, which need to occur, have occurred in the past, or exist in combination in order to cause harm. This is where other technology safeguards that act in association with, or in addition to, the safety-related control function are taken into account. A checklist of generic causes, to include operating mode and human factors, is provided to aid this process. Guidance is also given to explain how to handle common causes and the situation where there may be more than one possible scenario leading to the accident. The list of causes are recorded on form 2, as shown overleaf, in addition to the three mandatory ones.

Once a list of causes has been produced, a probability needs to be assigned to each cause. Guidance is given on how to estimate these probabilities with reference to the cause type classification taken from the checklist. Note that the probability of the safety-related control function failing, (cause 3) is fixed, at 1.0, because the purpose of the methodology is to calculate the amount of risk reduction that this function is required to provide and hence its SIL. Whilst aligning the work the operator will be in the danger zone 100% of the time, so the probability will be 1.0. However, for the accident to occur, the operation of the press would need to be initiated, for example, by pressing the foot pedal, whilst the hands were still in the press. The lookup table L3, shown below, giving the probability of human error when performing tasks under given constraints, is provided to enable the team to estimate the likelihood of this happening. This lookup table can also be used to estimate the likelihood of a specified functional test not being carried out. This information is recorded on form 2 as shown overleaf. Other lookup tables (not included in this paper) giving frequencies of rare events and probability values in terms of qualitative phrases are also provided for use with different types of causes.

Lookup Table L3 Probability of human error

Probability	Task
10^{-5} - 10^{-6}	Routine, good feedback with time to make use of it, good appreciation of hazard
0.001	Routine, simple
0.01	General error of omission
0.1	Non-routine, complicated
0.1	High stress, time constraint 30 minutes
0.9	High stress, time constraint 5 minutes
1	High stress, time constraint 1 minute
1	Error in second step, having already erred in first

All the probabilities are then multiplied together as instructed at the bottom of form 2. This value is then inserted into form 3 as shown in the example. The team is then required to take into account any possibilities to avoid or limit the harm. They do this by estimating a probability, for each of the four severity levels that could result as a consequence the accident scenario occurring. Examples of types of injury for each of these severity levels are given in another lookup table, L5, as shown below. The information in this table has been drawn from a number of international injury classification schemes [4-10]. The total probability for all four severities must add up to one. For example, with the power press, the team may consider that there is only a one in ten chance that the operator will be able to remove their hands quickly enough to avoid harm that if they don't it will result in an irreversible injury.

These values are then multiplied by the probability of the accident occurring in order to calculate the risk of each severity level in terms of the probability that this level of harm will occur.

These values are then used to calculate the required risk reduction (per hour) to be achieved by the safety-related control function, as shown in the fourth and final form. The most onerous of these values is selected and a final lookup table, L4 shown below, based on that given in IEC 61508 is used to obtain the target SIL.

Lookup Table L5 Severity level definitions

Severity level	Example injuries
Fatality and permanent serious disability	- Quadriplegia - Paraplegia - Prolonged unconsciousness (coma) - Permanent brain damage
Irreversible injury (major)	- Any fracture (other than to fingers, thumbs or toes) - Burns causing permanent scarring - Damage to sight partial or total - Any amputation - Loss of consciousness (not prolonged) - Dislocation of shoulder, hip, knee, spine - Treatment required due to fume exposure - Anything requiring resuscitation
Reversible injury (minor)	- Cuts and bruises - Minor burns, temporary scarring - Anything else requiring first aid only - Minor broken bones (fingers, toes)
No injury	

Lookup Table L4 SIL failure rates for high demand/continuously operating safety function

SIL	**Failure rate** (per hour)
4	$< 10^{-9} - 10^{-8}$
3	$< 10^{-8} - 10^{-7}$
2	$< 10^{-7} - 10^{-6}$
1	$< 10^{-6} - 10^{-5}$
0	$< 10^{-5} - 10^{-4}$
-	$> 10^{-4}$

The target SIL is based upon achieving a broadly acceptable risk for all of the three severity levels. The authors of this methodology propose that a broadly acceptable risk should be:

Risk of death: $1x10^{-6}$ per year

Risk of irreversible injury $1x10^{-5}$ per year

Risk of reversible injury $1x10^{-4}$ per year

The order of magnitude difference between the criteria for different severity levels helps the ease of use of the methodology. It can be justified by comparing the differentials in compensation payments made by various schemes such as: criminal compensation in the UK [10], worker compensation in Australia [8], and private accident insurance, for various injuries in each category. This order of magnitude is also implicit in various risk ranking techniques found in the literature [11-13].

This methodology for the risk-based allocation of SILs for safety-related control functions is only one input to the decision of the most appropriate SIL that the function is to be designed to. Other considerations such as reasonable practicality, societal concerns, current good practice and state of the art also need to be taken into account. It is expected that within the normative clauses of the standard clear guidance will be given as to when the SIL needs to be increased or when it can be relaxed. A higher SIL may be required, for example, due to societal concerns. There will need to be a clearly justified reason for relaxing the SIL from the target. One such justified measure can be by the use of additional non-E/E/PE safeguards. There will also be guidance as to when it is appropriate to use this methodology and when it is not. In fact, in the case of the power press example, because there is a recognized industry standard, it would be unnecessary to follow this methodology. Furthermore, this methodology cannot be used to justify straying from the requirements laid down in machine specific standards and codes of good practice.

CONCLUSIONS

This paper has described a methodology for the risk-based allocation of safety integrity levels (SILs) for safety-related electrical control systems of machinery. This methodology encourages the documentation of assumptions, takes into account the risk reduction provided by other technologies and promotes the discussion, within a team, about the interaction of safety-related control functions with other technologies and one another. This methodology is only one input to the decision as to the most appropriate SIL and is available for use when there are no machinery specific standards or codes of practice upon which to base this decision. Finally, there is much work in progress on all clauses of IEC 62061, including this methodology, so it is subject to change before the next CD is published at the end of 2001.

ACKNOWLEDGEMENTS

The authors would like to acknowledge, the contribution of the following people Tom Treble from HSE (part of the project team who developed the original methodology); Steve Frost also from HSE; Mark Charlwood from Innovation Electronics; Brian Clark OBE from BJC Enterprises and Derek Jones from EJA Limited. All of whom have contributed to the development and review of this work providing valuable information, ideas and comments.

REFERENCES

[1] IEC 62061 – Safety of machinery – Functional safety – Electrical, electronic and programmable electronic control systems, Committee Draft Sept. 2000

[2] IEC 61508:1998 – Functional safety of electrical/electronic /programmable electronic safety-related systems

[3] Hibbert, L., 1999, "Machinery Falling Short of the Mark", Professional Engineering, Vol. 11, No. 22

[4] National Safety Council, USA, "Classification of Motor Vehicle Traffic Accidents", 5th Ed, ANSI D16.1-1989

[5] Canadian Standards Association, "Coding of Work Injury or Disease Information", Z795-96,

[6] ILO, Geneva, 1976, "International Recommendations on Labour Statistics"

[7] Swedish "Injury Reporting Regulations"

[8] Australian "Workplace Injuries Compensation Guide"

[9] HSE, "UK Reporting of Injuries, Diseases and Dangerous Occurrences Regulations 1995"

[10] UK Home Office, "Compensating Victims of Violent Crime: Changes to the Criminal Injuries Compensation Scheme", HMSO

[11] Raafat, H., 1995, "Machinery Safety: The Risk Based Approach", Technical Communications (Publishing) Ltd.

[12] Allum, S. and Wells, 1993, G., "Short Cut Risk Assessment", Trans IchemE, Part B, Vol 71 161-168

[13] Warren, S. B. and Amundson, T. M., 1995 Comprehensive Baseline Hazard Assessments – A Team Approach", Professional Safety

Form 1 Define accident scenario

Definition	Description for machine
Potential accident Description of what has the potential to cause harm, what that harm is and how it is caused.	*Crushing / amputation of hands by press*
Intended operation Description of activity being carried out by machine and/or operator when accident could occur.	*Aligning work piece*
Safety-related control function Description of safety-related control function.	*To prevent the activation of the press whilst the operator hand's are in danger zone*

Form 2 Probability of potential accident

List of causes Use checklist to generate this and follow associated guidance.	**Type***	Id	Probability
1 Operator in danger zone Probability that person is exposed to hazard during any intended operation. Use Lookup Table L2.	NONE	p_1	1.0
2 Functional test on safety-related control function not carried out. What is probability that either no functional test is specified or it is omitted by the operator?	NONE	p_2	0.1
3 Safety-related control function fails This is the control function for which a SIL is being decided. Failure probability fixed.	FIXED	p_3	**1**
4 Foot pedal pressed before hands removed	HUMAN	p_4	0.1
5 etc. etc.		p_5	
Probability control function failure results in potential accident $P_1 = p_0 \cdot p_1 \cdot p_2 \cdot p_3 \cdot p_4 \cdot p_5 \cdots$	NONE	P_1	0.01

Form 3 Probability of different severity levels

Component			**Identifier**	**Value**
Probability of accident (from Form 2)			P_1	0.01
Probability that if accident occurs it results in a particular level of harm			**Risk**	
Severity level (see Lookup Table L5)	**Id**	**Value**	**Identifier**	**Value**
Fatal and permanent serious disability.	p_f		$P_4 = P_1 \times p_f$	
Irreversible injury	p_i	1.0	$P_5 = P_1 \times p_i$	0.01
Reversible injury	p_r		$P_6 = P_1 \times p_r$	
No injury	p_n			
Total		**1**		

Form 4 Required risk reduction

Severity level	**Required risk reduction (per hour)**	
	identifier	**value**
Fatal and permanent serious disability.	$R_1 = 10^{-10}/P_4$	
Irreversible injury	$R_2 = 10^{-9}/P_5$	1×10^{-7}
Reversible injury	$R_3 = 10^{-8}/P_6$	
Most onerous required risk reduction (per hour)	lowest of R_1, R_2, R_3	1×10^{-7}
Required SIL (use Table L4)		SIL 3

Proceedings of
2001 ASME International Mechanical Engineering Congress and Exposition
November 11–16, 2001, New York, NY
SERA-Vol. 11

IMECE2001/SERA-24007

PRINCIPLES FOR PROCEDURE SYSTEM MANAGEMENT

Maggie Sturdivant and Earl Carnes
U.S. Department of Energy
19001 Germantown Road • Germantown, MD 20874 USA
(301) 903-0077 (voice)• (301) 903-6172 (facsimile)
maggie.sturdivant@eh.doe.gov or earl.carnes@eh.doe.gov

ABSTRACT

This paper provides a preview of a DOE Technical Standard "Principles for Procedure System Management" due to be released in the fall of 2001. Procedures are identified as causal factors in accidents, operating events, and lessons learned within the U.S. Department of Energy (DOE), the commercial nuclear industry, and other complex industries. While literature exists regarding the control, production, and use of procedures, DOE has not provided guidance on the collective elements and attributes of an effective procedure system. This paper describes a set of design control features that can systematically improve development and maintenance of procedures.

INTRODUCTION

A previous paper, "Knowledge Transfer for Emerging Work Paradigms," presented at the 2000 ASME Congress summarized the work of the DOE Procedures Topical Committee to develop the basis for a Procedure System standard, a principles-based enterprise approach for knowledge transfer. Members of the Committee include managers, operators, scientists, engineers, and technical specialists who have years of experience with procedures and management systems as well as with design and operation of DOE's unique facilities.

The 2000 paper focused on experiences in DOE that led to formation of the Committee, summarized relevant research, and described the basis for the proposed standard. The Committee noted an increasing emphasis in regulations and standards on the importance of procedures. Examples include: DOE's Quality Assurance Requirements contained in 10 CFR 830 Nuclear Safety Management; ISO 9001 Quality Management Systems - Requirements; 10 CFR 1910.119 Process Safety Management of Hazardous Chemicals; and 40 CFR Part 68 Risk Management Programs for Chemical Accidental Releases. However guidance had not been provided on systematic methods for developing and managing procedure programs. At the time of the Committee's inquiries DOE contractors were implementing Integrated Safety Management (ISM) Systems. The Department's ISM policy and requirements were designed to establish a unifying framework and a systems approach whereby all DOE contractors would include design for safety as an integral part of work design. Within this context the standard on procedures establishes a systems approach and sets forth the design features for a procedure system - the principles of the standard. These design features are the focus of this paper.

The DOE standard presents guidance for integrating worker experience, management and technical knowledge, commitments, requirements, and standards into the documented procedures that effectively support the missions of the DOE. It was designed as an enterprise-level guidance document applicable to all DOE work — scientific research, environmental cleanup, national security, energy supply, and administration. It is written from a system perspective and intended to be of use to individuals who have procedure system management responsibilities. The guidance is expressed as statements of principles which make evident the logic of a procedure system and constitute the design control features of a system and the procedures it produces.

The work, hazards, and environment in which the work occurs and the skills, knowledge, and experience that the work demands give rise to the types of procedures to be used and how procedures will be used in performing work. Procedures specify and describe the direction for how administrative and technical work are to be accomplished. They also convey management expectations for the decisions and actions of individual work performers. Procedures are products of cross-functional collaboration, capturing and integrating knowledge and experience from multiple disciplines. They communicate to the performer of work the organization's collective knowledge of how to perform work while protecting the health and safety of people and the environment. A procedure system built on design control features amplifies and channels the collective efforts and expertise of line management, workers, and subject matter experts to produce procedures that are correct and usable.

PROCEDURE SYSTEMS

A procedure system is a fundamental part of an organization's integrated management system. It is the infrastructure that makes possible development and maintenance of reliable procedures. An effective procedure system produces work-focused, accurate, and usable procedures.

The term "procedure system" describes a distributed set of systems and processes (Figure 1). Such a system integrates input for performance of work communicated from other management systems. Procedure system inputs include work processes, activity and task analyses, hazards analyses, safety analysis documentation, work task analyses and human performance analyses, design documents, specifications, drawings, calculations, environmental controls, authorization bases, authorization agreements, and safe work practices.

The procedure system processes include development, review, approval, distribution, use, maintenance, cancellation, retirement, and archival of procedures. Line management responsibility and accountability for overall procedure performance and compliance should be assigned. Because the procedure system has elements distributed throughout the organization, authority and accountability for each procedure system element should be described. This assignment of authority and accountability should ensure availability of resources — funding, personnel, data management, technical expertise (subject matter experts; technical and safety reviewers) — and associated production resources. It should also ensure that procedure-related feedback, operating experience, and lessons learned are directed to the procedure system

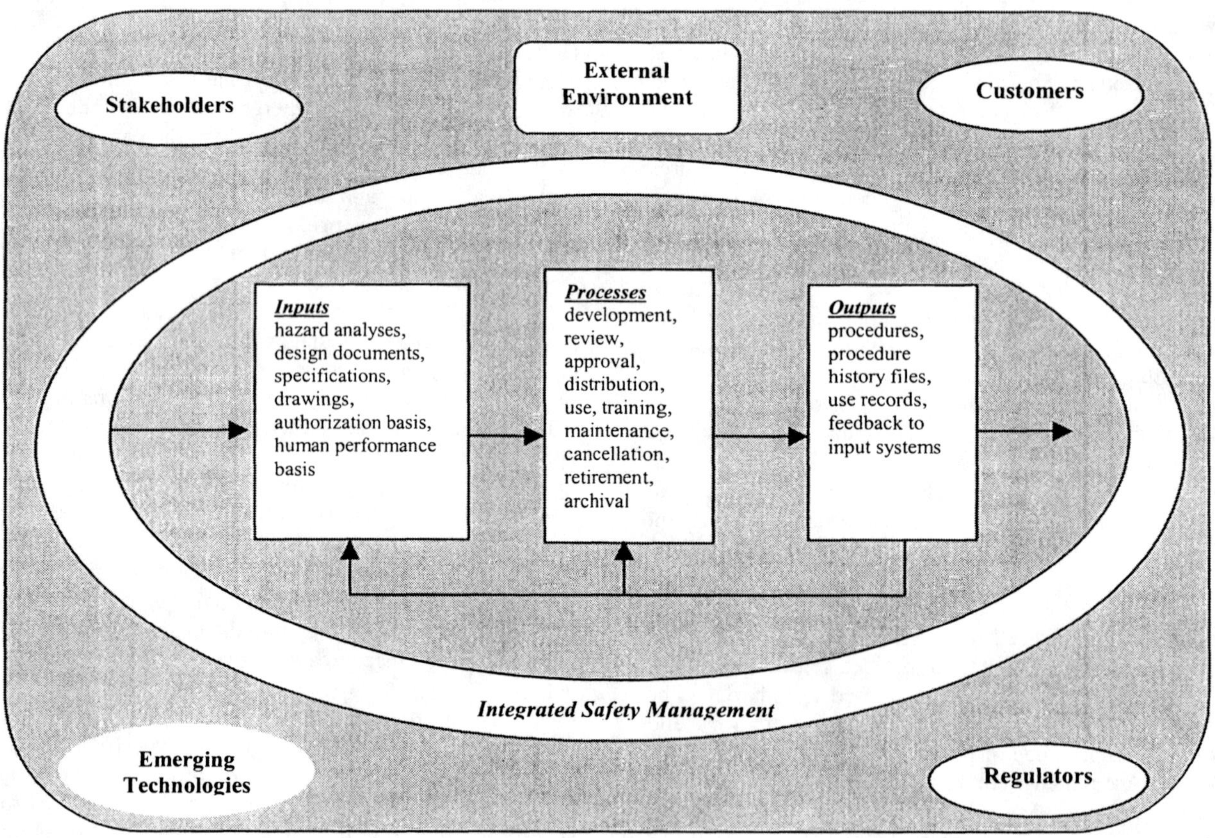

Figure 1: The Procedure System and Work Environment

The products of procedure systems are the work control directions (management controls and technical procedures) used by administrators, operators, supervisors, craft, scientists, engineers, and other workers during the performance of work. Outputs also include procedure history files, use records and feedback to input systems. The quality and usability of the procedures for working effectively, efficiently, and safely are dependent on the reliability of the procedure system, the distributed elements of which are described as part of the responsible organizations' formal management controls.

Procedure systems govern two broad categories of procedures — management control for administrative functions and technical procedures.

Management Controls establish the set of management functions (such as quality assurance, research and development, and finance) necessary to perform an organization's mission safely, effectively, and efficiently. Management controls are derived from an analysis of mission requirements at each activity level and are drawn from regulations, compliance agreements, contracts, professional standards, codes of practice, organizational policy, management philosophy of operation, and best practices. Implemented as intended, they shape much of the "culture" of the organization.

Technical procedures are used primarily for facility, activity, and task work. They are based on design controls (specifications, drawings), derived operational controls (hazards analyses, documented safety analysis reports, and technical specifications), work area management controls (industrial safety, training), and operating experience (feedback and lessons learned programs).

Design Features For Procedure Systems

Procedure systems tailor procedures to fit the work. Tailoring emphasizes the need to balance priorities and to fit work controls to the work. This is a deliberative and rigorous exercise requiring skills, experience, and knowledge of the work, its hazards, and the work place. Tailoring is central to setting design features for effective procedure system management.

A procedure system requires the following design features (DFs):

DF1: Criteria identify the work for which procedures are needed.

Defining the work is the starting point for determining the procedural controls necessary to support work execution. Defining the scope of work consists of translating mission program or project objectives into work controls; eliminating, reducing, or mitigating hazards; setting performance expectations; identifying and prioritizing tasks; and allocating resources. Work controls include engineered design features, administrative controls, work practices, personnel protective equipment, and skill of the trade. Administrative controls and work practices should be developed and maintained though procedure systems.

Work must be designed to be safely doable if procedures are to aid and not hamper disciplined and effective performance. Critical work design inputs that are routinely examined to determine the control set for a given work scope should include:

- ✓ Degree of complexity of the work;
- ✓ Hazards associated with process, activity, and/or tasks;
- ✓ Degree of uncertainty associated with the work or hazards;
- ✓ Level of risk to be accepted;
- ✓ Skills, knowledge, experience, and training of work performers;
- ✓ Duration of the work (short-term activity or ongoing project);
- ✓ One-of-a-kind or recurring work activity;
- ✓ Whether work takes place in multiple or single locations;
- ✓ Degree of decision making granted to work performer.

Procedure use should be stipulated to support risk reduction and to prescribe a standardized way of doing work where substantial consistency is important to safety, quality, human performance, or reliability or where consistency is essential to maintaining compliance with laws, regulations, and commitments.

DF2: Competencies, roles, and authority are clearly defined for individuals who participate in procedure system processes.

The organization should establish provisions to determine that managers, procedure developers, procedure reviewers, verifiers, validators, and procedure users are competent commensurate with their responsibilities. Competence factors include knowledge and experience, the level of supervision involved in assigned tasks, and whether assigned tasks will be performed individually or by a team.

All individuals involved in procedure development, review, verification, and validation should be trained to understand the procedure system. Procedure developers should be trained in performing work and human performance task analyses tailored to the complexity and uncertainty of the work, developing and documenting the procedure bases, and using writers' guides. Reviewers, verifiers, and validators should be trained to perform appropriate technical and interdisciplinary reviews. Verifiers and validators should be trained in the methods and requirements of verification and validation. Procedure users should be trained on procedures that govern the specific work they perform, policies on procedure use, and operation of the procedure system.

Training should emphasize that procedures are the responsibility of the managers and workers who are accountable for proper development, production, and use of procedures.

DF3: Mechanisms are identified for analysis of the work and expected hazards as the basis for work controls selection.

As a fundamental part of an organization's integrated management system the procedure system has inherent objectives to avoid unwanted or harmful outcomes and to minimize delay in the achievement of mission objectives. A procedure system relies on other management systems and processes to provide safety and other risk management bases information to use in procedure development. Organizations have specific programs for analyzing work processes in order to identify hazards and other success barriers in the work environment. These analysis mechanisms are governed by requirements and standards and generate products that provide basis information for developing procedures. The quality of the analytical inputs provided from these systems directly affects the quality of the procedures. Requirements should be established to ensure that all sanctioned sources of risk and hazard analysis are available and comprehensible to those responsible for procedure development.

DF4: Human performance analysis examines the interactions of people with the systems and environments in which they work.

Organizational factors, human-systems interfaces, training, shift working hours, time pressures, and hazardous work environments all affect human performance. The content, form, and method of delivering procedures to the worker should be formulated based on a careful analysis of the nature of the work, hazards, and the conditions within which the work will be performed. Work must be designed to be safely doable if procedures are to aid, and not hamper, disciplined and effective performance.

Work process, activity, and task analyses are essential inputs to a procedure system. These analyses examine work sequences — what initiates the action; the steps to be followed in response to that action; and the results from completing those steps. The analyses identify potential hazards, vulnerabilities, or the areas that increase the probability of errors, and the consequences of these errors. The intent of such analyses is to produce a set of directions and information to be addressed in the procedures and to identify human performance issues, indicate the appropriate warnings and cautions, and determine qualification or training necessary for persons who will carry out the procedure.

For Management Control procedures, work process analyses often focus on maximizing efficiency, cost-effectiveness, productivity, and communication.

For technical procedures, factors to be evaluated include:

- Environmental conditions, such as noise, temperature, confined spaces or hostile breathing atmosphere;
- Use of protective clothing or equipment;
- Physical manipulation of equipment;
- Visual or audio feedback from equipment manipulation;
- Task performance by individuals or teams;
- Worker familiarity with tasks, equipment, and work location;
- Time critical factors;
- Availability of external support (e.g., radio contact with control center, ability to stop work and seek support);
- Degree of precision required (e.g., set to a precise temperature or within a range of temperatures).

Efforts should be made to identify ways in which errors could be made in task execution and to design the procedure content, form, and delivery to prevent or mitigate known and anticipated error types. Procedure design methods should be applied for eliminating common errors of commission and omission. Design issues that can be addressed include:

- Common error types — knowledge of common errors made when using procedures and strategies for prevention;
- Selection of presentation media — electronic media, paper copies;
- Presentation layout — how to structure the presentation to increase comprehension and usability;
- Narrative layout — how to use headings, white space, and typographic techniques to increase performance and quality;
- Language — how to use terminology, reading levels and sentence structure to enhance understanding and usability;
- Graphical presentations — how to use non-narrative formats (such as tables, figures, graphs, and forms) to depict required actions and documentation;
- Combined presentation structures —narrative descriptions along with job performance aids or postings.

DF5: Approved methods describe assembly of adequate technical and administrative bases to guide procedure development.

All procedures, whether management control or technical, should have an identifiable basis. The procedure basis should describe how the work controls selected (1) can be identified with the performance expectations of responsible management and (2) can be shown to be tailored to the specific challenges and hazards presented by the work. A procedure basis consists of a complete listing of source documents referenced to develop a procedure, a description of how these source documents were selected, and a description of assumptions and informed judgments used to integrate requirements and supporting information from the source documents into the procedure.

Source documents vary depending on procedure type. For technical procedures, hazards analyses and design specifications are principal information sources. For management control procedures, source documents include policies, regulations, contracts, standards, and best practices.

DF6: Interfaces with other procedures are identified and linkages are established.

Understanding and maintaining the configuration of the total set of procedures is as important as understanding and maintaining the configuration of systems, components, and equipment. Introducing a new procedure or a change to a procedure can introduce new conditions that would affect other procedures.

The processes for developing and revising procedures should identify both the interfaces with other procedures and a means of change control that maintains the configuration of the complete procedure set. Guidelines should specify how procedure interfaces among related sets of procedures are established and maintained.

DF7: Verification of accuracy occurs through formal interdisciplinary reviews.

Verification substantiates a procedure's technical accuracy. The verification process is intended to ensure that:

- The procedure is technically accurate;
- The development and review processes have been conducted consistent with the applicable management controls;
- The procedure is consistent with existing related procedures and consistent with ongoing initiatives;
- The development and review processes have not inadvertently introduced requirements for actions that are inconsistent with contracts and approved operating practices.

The skills, knowledge, and experience mixture applied for verification should be comparable to that applied to developing a procedure. The procedure review process draws on relevant experience and competence and facilitates the interfaces of the procedure with other activities. The type and depth of verification depend on the procedure content, complexity of the activity, levels of uncertainty, operational and safety considerations, and degree of coordination needed among programs and areas of expertise.

DF8: Validation for usability occurs through walk-through or comparable methods.

Validation is conducted prior to approval to ensure that the procedure can be used as written. Validation allows independent assessment after resolution of review comments to demonstrate that a procedure can be used as written in the work environment where the actual task is to be performed.

All procedures are validated. The level of confidence needed depends on several factors:

- Type of procedure;
- Types of work and the degree of hazard;
- Operating experience or the degree of familiarity with the operation (e.g., a small change to a well-documented and mature process as contrasted with work involving minimally characterized hazards, high uncertainty, or high complexity);
- Successful work experience, including lessons learned materials and other supporting documentation.

For management control procedures, validation may be completed as part of the technical and interdisciplinary reviews of the procedure.

For technical procedures, some level of simulation of the actual practice is used. At the highest level of confidence, validation involves use of a mock-up or simulator in which the actual work steps are practiced. At a second confidence level is a walk-down in which personnel take the procedure to the location where the task is to be done and, without actually performing the task, ensure that each work step is correct, the procedure steps can be performed as written, and the entire work scope can be performed as directed in the procedure. At a third level of confidence, validation is an analytical review that does not simulate actual operation, but validates the procedure based on the knowledge and experience of the reviewers.

DF9: Formal procedure approval establishes accountability commensurate with authority and responsibility of affected managers, supervisors, and workers.

A procedure should be approved by a person with authority to direct conformance with the procedure. This approval should attest that the procedure is technically correct, that workers can understand and use the procedure, that appropriate verification and validation activities have been completed, and that the new procedure or a change to an existing procedure has been reviewed for impact on related procedures. Approval for use of a given procedure should be based on a determination that related work would not be adversely affected by implementing the new or revised procedure.

DF10: Criteria identify the interface between procedure approval and authorization of procedure use.

The approval process should require the approval authority to determine the implementing requirements associated with authorizing use of a procedure. These implementing requirements include, but are not limited to the following:

- Resources should be adequate to implement the procedure. Resources include personnel, hardware, and associated equipment.
- Status of related procedures, including development or revision of other procedures that interrelate with the procedure to be approved.
- Training needed for workers who will use the procedure. Depending on the complexity of the work and the experience and qualification of personnel, training may range from reading the procedure, to reviewing it with supervisors, to formal training.
- Management and technical interfaces are satisfied.

DF11: Revision controls and performance evaluation requirements preserve the integrity of procedures.

Maintaining the integrity of procedures and the configuration of the procedure system requires a robust change control process. Evaluation and periodic revision of procedures are necessary, ongoing processes required to keep procedures current with changes in mission, work, hazards, requirements, systems, personnel, and equipment. In addition to supporting routine changes introduced as part of regular operations, the revision process should also include a process for expedited review and approval of procedure changes when urgency demands immediate resolution. A method for canceling existing procedures should be established to ensure that procedures that are no longer necessary or no longer used are effectively removed from the system.

Procedure systems should account for the following factors that typically drive revisions to procedures:

- When work changes;
- When missions are substantially modified;
- When there are changes to the authorization basis, technical basis, facility configuration, standards, commitments, or other bases;
- When there are unusual operating conditions and configurations;
- When there are evaluations, such as user feedback, periodic review, lessons learned, self-assessment, and audits;
- When there are identified opportunities to improve human performance;
- When there are changes in requirements, commitments, or expectations;
- When there are identified procedure deficiencies (errors, omissions);
- When there are identified Unreviewed Safety Questions;
- When work force qualifications undergo significant change;
- When contract changes occur.

DF12: Mechanisms are identified for maintenance and retention of procedures and supporting documentation.

An important part of the overall procedure system is controlled delivery to ensure that the correct and current version of a procedure is available where and when needed. A delivery system commensurate with the volume and types of procedures provides a means to ensure that procedures are in place, are made available to procedure users, and the procedure users know where to find them. All procedure presentation types — hard copy, electronic, and user aids — must be controlled in accordance with procedure system requirements. As appropriate, the delivery system should accommodate:

- Procedures to be issued on a routine basis;
- Procedures that require expedited processing;
- Procedures that may require special controls due to sensitive, proprietary, or classified information.

Ready access to the history of proceduralized operations is needed for development of new procedures, revision of existing procedures, procedure reviews, and audits. The records program should include access to and retrievability of procedure records. The records program must:

- Retain procedure files and procedure-generated records, such as checklists, forms, and logs, as part of an organization's records programs;
- Maintain results of periodic procedure reviews and audits to document that procedures were reviewed on a regularly scheduled basis for technical accuracy and usability;
- Afford timely and effective means to retrieve superseded procedures and interrelated supporting materials (e.g. basis information);
- Retain and archive records generated through procedure use.

DF13: A method of feedback is identified for continuing improvement of individual procedures and of the procedure development process.

Mechanisms should be established to collect data on the effectiveness of procedures during work performance, whether one-time use or multiple use. Items that should be evaluated as part of procedure use include:

- Did the procedure accurately address hazards and protective measures associated with the work?
- Did the procedure identify the necessary equipment and material to support the work?
- Did the procedure steps correctly correspond to the task to be performed?
- Did the procedure correctly identify interfaces with other work activities, other procedures and other work controls?
- Was the procedure structured to be easily understandable and useable for the procedure users (e.g., was there too much branching within the procedure or referral to supporting documents)?
- Did the procedure allow users to easily record data to be collected during the work activity?
- Did the procedure contain extraneous or unnecessary information?

The procedure system should demonstrate how feedback and continuous improvement are used throughout the procedure development process to iteratively perform quality checks in order to consistently produce effective procedures. Opportunities for improvement and correction should be identified for each step in the procedure system process, beginning with identification that a procedure is required through the development, review, and approval of the procedure and the performance of the work. Any opportunities for improvement or correction should be documented though the appropriate quality and corrective action programs as well as the history files of the individual procedures involved.

Procedure system requirements should ensure that lessons learned from procedure development and use are recorded and reflected, as appropriate, in enhancements to the individual procedures; the procedure development process; or related programs, such as personnel qualification and training. The careful evaluation of basis inputs should be used as an opportunity to feedback potential improvements to bases documents and the systems and processes that deliver those documents. Post-job briefings and higher level reviews can capture experience in using procedures immediately following work performance or an independent assessment. Critiques by procedure developers and reviewers can aid in capturing information about the procedure development process. Self-assessments, independent assessments by DOE or independent reviewers, benchmarking other organizations, and evaluation of best practice reports provide valuable input for improvements and corrective actions.

CONCLUSION

The DOE work portfolio includes scientific research and development, energy supply, nuclear weapons stockpile and stewardship, and cleanup of the environmental legacy from weapons production. How can a large and complex organization provide value-added guidance on the development and maintenance of procedures? The DOE Procedures Topical Committee's evaluation of procedure problems within DOE concluded that a systems approach was required.

The Committee identified design features that bounded procedure problems experienced across all mission work. These have been expressed in a DOE standard, "Principles for Procedure System Management," as statements of principles with enterprise applicability. The statements of principles are organized into three categories. The first category provides guidance on institutionalizing a procedure system though policy and assignment of responsibilities. The second addresses development, approval and maintenance of procedures. The third offers guidance on supporting a procedure system through change control, document control, records, information management and training. Collectively the principles make evident the logic of a procedure system and constitute design control of the system and the procedures it produces.

REFERENCES

Sturdivant, M. H. and Carnes, W. E., 2000, *Knowledge Transfer for Emerging Work Paradigms*, 2000 ASME International Mechanical Engineering Congress and Exposition, Safety Engineering and Risk Analysis, American Society of Mechanical Engineers.

ACKNOWLEDGEMENTS

The authors wish to acknowledge Dr. Fred Carlson, F.N. Carlson and Associates, Inc., Dr. Daryl Grider, West Virginia State College, Institute, West Virginia, and Dr. Daniel Plung, Westinghouse Savannah River Company for their contributions to this paper.

Proceedings of
2001 ASME International Mechanical Engineering Congress and Exposition
November 11–16, 2001, New York, NY
SERA-Vol. 11

IMECE2001/SERA-24008

ROBUST DESIGNS, UTILIZING SIX-SIGMA QUALITY PROCESSES, LINKED TO A USMC WEAPON SYSTEM

Nick Strifas, Bob Kalamajka and Massimo Ruzzene*

NAVSEA Dahlgren Division
Code G53, Systems Engineering Integration Branch
Fax: 540-653-7440 Phone: 540-653-3028
strifasne@nswc.navy.mil

*The Catholic University of America
Department of Mechanical Engineering
Washington D.C. 20064

Abstract

This study describes the approach to integrate Six-sigma and Robust Design approaches on a US Marine Corps (USMC) program. Six Sigma and Robust Design techniques are tightly coupled to ensure that a design is both manufacturable and also serves the needs of the US Marines. Six Sigma is an aggressive (stretch) goal that sets a high standard of excellence for the development of a world class quality product (Short Range Anti-tank Weapon - SRAW) for the US Marines. This study also describes how the design team is linking Six Sigma and Robust Design, to ensure that a very complex product, with many steps and processes involved in its manufacture, can be recognized as a superior one.

Introduction

Six Sigma is basically thought of as a process quality goal. However, it must always be remembered that quality processes are one of the key elements of product reliability. Design for reliability encompasses a great deal more than this. If the individual parts themselves do not start with an adequate level of quality and reliable design is not adequately addressed during development, final product reliability at the desired level (at the six sigma level) is simply not possible. While building on Statistical Process Control (SPC) as its foundation, an effective Six Sigma program actually represents a paradigm shift. It goes far beyond just measuring, identifying and segregating defective items; it entails an entirely new management approach. It requires trusting all of the members of the workforce, and educating, and training them (including top management). The intent is a participative-management process, that emphasizes employee involvement at all levels. The program goal is teamwork. The primary goal of this approach should be to assure total customer satisfaction (improving quality and reliability). Understanding the customer, his needs and desires, and improving reliability are the focus of the effort.

As a normal result of the manufacturing process, all manufactured items are subject to item-to-item variation. This variation is the enemy of quality. Historically, controlling a manufacturing process to a nominal -3σ to $+3\sigma$ variation was considered adequate. This resulted in 99.73% of the items being within specification or conversely, 0.27% out of spec. The out-of-spec items were then (hopefully) removed following inspection. This level of defects, however, is nowhere near acceptable with today's highly complex equipment in our worldwide, highly competitive and quality-conscious marketplace. The Six Sigma process, which is based upon Statistical Process Control (SPC), carries the concept of high quality a great deal further. A manufacturing process actually controlled from -6σ to $+6\sigma$ (a total spread of 12 standard deviations) would assure that 99.9999998% of the items were within specification. Assuming a normal or Gaussian distribution (which is realistic for the typical parameters of interest), this represents seven orders of magnitude improvement [1].

Six-Sigma an Overview

Six-sigma is the benchmark of excellence for product and process quality, popularized by Motorola. It provides a quantitative, statistical notion of quality useful in understanding, measuring, and reducing variation. A product is said to be Six Sigma quality if there are no more that 3.4 non- conformities per million opportunities (3.4 parts per million) at the part and process-step level, in the presence of typical sources of variation. The Six Sigma quality concept recognizes that variations are inevitable due to insufficient design margin, inadequate process control, imperfect parts, imperfect materials, fluctuation in environmental conditions, operator variations, etc. Essential SPC metrics associated with Six Sigma programs are based on Six Sigma characterization of products and processes. The process capability (C_p) and the process performance (C_{pk}) are used as the vehicles to characterize the product-process quality.

Process Capability (C_p) is a measure of the ability of a process to produce acceptable products. C_p can be considered to be a short term metric. It is measured by comparing the variation in the product to the upper and lower specification limits, assuming that the product mean value is centered between the upper and lower specification limits. A high C_p (e.g., $C_p = 2$) indicated that the process is under control and capable of reproducing the desired characteristic. It makes no statement about whether the process is centered on desired mean [2].

$$C_p = \frac{|USL - LSL|}{6 \cdot \sigma}$$

where:
USL = Upper Specification Limit
LSL = Lower Specification Limit

Process performance (C_{pk}) combines both variance and off-target process means. C_{pk} can be considered to be a long term metric, i.e., it also accounts for process drift. A high C_{pk} indicates that the process is actually reproducing the desired characteristic within the desired limits.

$$C_{pk} = Minimum\_of\_\left[\frac{Target - LSL}{3 \cdot \sigma}; \frac{USL - Target}{3 \cdot \sigma}\right]$$

From this equation, we can see that when C_{pk} and C_p are identical the process mean is on target. However, actual process performance is somewhat degraded from the optimum, i.e., that the actual mean of the manufacturing process has been displaced somewhat from the process nominal specification target (typically a 1.5-sigma shift as shown in figure 1). Thus, when the process mean is 1.5σ off target, the shortest distance from the process mean to either specification limit is 4.5σ. C_{pk} is reduced from 2.0 to 1.5. Thus while a theoretical Six Sigma program would appear to require a $C_{pk} = 2.0$, in practice a $C_{pk} = 1.5$ is more frequently used. In spite of the degradation of process performance resulting from 1.5σ shift of the process mean, the process is still capable of producing items with an equivalent defect rate (e.g., + or - 4.5σ) much lower than the historical (3.0 σ). While these techniques were developed in a manufacturing environment, it is also important to understand that the application of Six Sigma program is not limited to the manufacturing arena [3].

Approach to Robust Design and Six-Sigma Program

Robust designs are characterized by sensitivity to variations in parts and materials, manufacturing process, and the operating environment. Design robustness is impacted by the flowdown of system requirements. Proper system requirements flowdown establishes initial allocations, margins, and tolerances.

Because of its global impact on the system, requirements flowdown is typically supported by techniques that enable design margin analysis. A robust design can be achieved by simultaneously minimizing design sensitivity and process variation. Depending upon the design phase, process variation and design sensitivity may be uncovered through analysis procedures such as design of experiments, simulation, parametric analysis, and SPC. The objective during design is to predict design sensitivity and link it to process variability. Efforts to reduce both design sensitivity and process variability should initially focus on design/process decisions that have the greatest impact on customer functional requirements. The first step in Six Sigma process is therefore to completely explore, define, and document system (customer) requirements. Reliability design tools are also considered in the design effort due to their potential impact upon design margins. These could include Failure Mode Effects and Criticality Analysis (FMECA), Fault Tree Analysis (FTA), Sneak Circuit Analysis, Worst Case Circuit Analysis (WCCA), Accelerated Testing. The basic premise is that it is much easier, and more cost effective, to eliminate bias from the design target characteristics than it is to reduce the variation in hardware. Thus the second step in S^4 is to allocate critical requirements traits to an appropriate level.

Robust design can be linked to Six Sigma through a defect classification system where total process characterization is the key to adequate process control. This involves establishing specific values of C_p and C_{pk} for each of the critical parameters and then optimizing the manufacture process. A closed-loop Failure Reporting, Analysis and Corrective Action System (FRACAS) is probably the most effective tool to help identify and correct these problems.

The implementation of Six Sigma program requires close cooperation between the product design effort and the design of the manufacturing process. Figure 2. shows the six steps approach, which was developed based on the experience and lessons learned from a US Marine Corps program, to achieve Six Sigma quality levels. The six steps process begins by identifying the physical and functional requirements for the product. Next, analysis is conducted to determine the characteristics of the product that are critical to accomplish each functional requirement. The third step involves further analysis to determine if each characteristic is controlled by parts, processes, or a combination of both. In the fourth step, the maximum range of each characteristic is explored and explicitly stated. During the fifth step, process variation that applies to each characteristic is determined using statistical quality control, design of experiments, and other applicable techniques. Finally, the results of the preceding five steps are combined during the sixth step to enact design improvements through the redesign of materials, assemblies, components, and processes. The goal of the six step process is to refine and improve designs (and the documentation of those designs) to the point where they have reached Six Sigma manufacturability levels [3].

Six-Sigma A US Marine Corps Program

The real value of a Six Sigma program is the process of continuous improvement that produces good results despite the possible shortcomings of some of its metrics (i.e., C_p & C_{pk}, also defect tracking & charting cost growth). The Short Range Anti-Tank Weapon (SRAW) system final design bears only a superficial resemblance to the prototype missiles flown during the demonstration phase of the program. Each component of the system has been subjected to intense scrutiny to assure that factors of performance, reliability, weight, and cost were allocated in a rational manner. The disciplined engineering design approach included provisions for on going producability analysis, process design, manufacturing planning and technology development, and tooling and inspection/test equipment design. The same approach was applied to the allocation of inspection and test criteria into the design. There has been continuous emphasis on inspection and test on the lowest possible level, in order to reduce defect rates at higher, greater value added levels. This emphasis has continued throughout the mechanical and electronic producability design evolution. Such innovations as using integrated plastic parts to replace metal parts and fasteners, reducing control components, using molded parts and rigid flex boards to replace machined parts and wire harnesses, use of bonded joints, utilizing multiple rigid flex circuit card assemblies, integration of in-circuit test into assembly operations, minimizing number of connectors, use of Application Specific Integrated Circuits (ASICs), and performance of environmental stress screening at the lowest possible level of assembly have helped reduce parts count from 1500 during the demonstration phase to approximately 500 in the final design. When the parts count goes down, so does the amount of inspection and test necessary. Throughout this design effort, parts and assemblies have been subjected to inspection and test in order to assure their performance and characteristics conform to the six-sigma design guidelines. As the design was approaching its Product level of maturity, these studies were utilized to reestablish the tests, inspections, and the acceptance criteria in a closed loop process.

Tracking Ongoing Product Improvements

Process control charts are useful to monitor the process stability and identify when special situations occur. A process is considered to be " in control " whenever the process and the measurements are within the upper control limit (UCL) and Lower Control Limit (LCL), which are positioned around a center line (CL). These control limits should be independent of any specification limits. Process control charts were

Utilized during the EMD (Engineering & Manufacturing Development) phase in the SRAW program to monitor individual modules and make sure that their processes were in control. A tracking chart of the reduction of defective components is shown in figure 3.

Improving Product Development (Lessons Learned)

The following suggests that statistical techniques could be a more integral part of the product development process, not just used to identify what should be done different to improve the process. The formal and disciplined implementation suggests improvements to the model used on the Marine Corps Program. The intent is to stimulate thoughts on how to create a project within the development process (this is where the six sigma benefits can be the greatest).

1. *Provide education in statistical methodologies.* The fundamental methodologies of statistical process control design of experiments (DOE), brainstorming, TAAF (Test, Analyzed And Fix), and quality function deployment (QFD)-type activity must be used by large numbers of individuals to develop, manufacture, and deliver the best possible product.
2. *Identify and optimize essential process and product parameters.* Defining all processes and creating flow charts that describe the steps of the process can give enlightening insight that may not be otherwise apparent. The list of processes needs to include existing basic internal processes in the development process. Brainstorm sessions that bridge departmental barriers may indicate the changes that are needed to improve the process so that it better meets the overall needs. Establish processes that identify the parameters that affect product performance and quality early in the design and development phase. DOE response surface experiments, and pass/fail functional testing are tools that need to be considered as an integral part of these processes.
3. *Define tolerances on key perimeters.* QFD is a tool that can be helpful when defining limits relative to meeting the needs of the product. DOE and statistical tolerancing techniques can also be useful tools to determine tolerance levels.
4. *Construct control charts, establish control limits and determine process capability indices.* The initial identification and tracking procedures for key parameters need to begin in development. Continuous response

outputs should be used wherever possible. The initial data collection should begin as part of the development process.

5. *Implement statistical process controls in the development line with a management system to assure compliance.* Functional statistical process control is a tool that can monitor the key product parameters as a function of time. An implementation process is needed to address both sporadic/special (i.e., out-of control) conditions and chronic/common conditions (e.g., the number of defects is consistent from day-to-day; however, the amount of defects is too high). An implementation process must be created that encourages continual process improvement, as opposed to "fixing all the problems" via "fire-fighting"
6. *Demonstrate process capability indices for key processes.* For an accurate quantification of common-cause variability, the process capability indices need to be calculated from data that were taken from a stable process over time. It is beneficial to make an estimate of the process capabilities for key parameters and their tolerances with early data. Results from this activity can yield a priority list of work that needs to be done with a starting point where robust designed experiments (i.e., DOE strategy) are used for optimization and the reduction or product variability. To meet variability reduction needs, it may be necessary in some cases to change the design or process.
7. *Transfer the responsibility for continuous process improvement to manufacturing.* The manufacturing organization should be involved within development activities. SPC charting needs should be conducted on all key parameters with an emphasis on making continual process improvements where required.

Conclusions

The goal of the six step process is to refine and improve designs (and the documentation of those designs) to the point where they have reached Six Sigma manufacturability levels. The Six Steps to Six Sigma Manufactured Products process is based on product characteristics that are critical to physical and functional requirements and to product elements that contribute to achieving these critical characteristics. In this study we have shown how the Motorola Six Sigma approach can be integrated into a very complex product providing continuous improvement and excellent results.

Acknowledgments

The authors would like to thank SRAW's Technical Direction Agent Mr. Michael Block for his support.

References

1. START, "Selected Topics in Assurance Related Technologies," Reliability Analysis Center, Volume 6, Number 5, 2000.
2. M. Harry, "Six Sigma Mechanical Design Tolerancing," Motorola Technical Press, 1988.
3. Forresr W. Breyfogle III, "Implementing Six Sigma," John Wiley & Son, Inc., New York, NY, 1999.

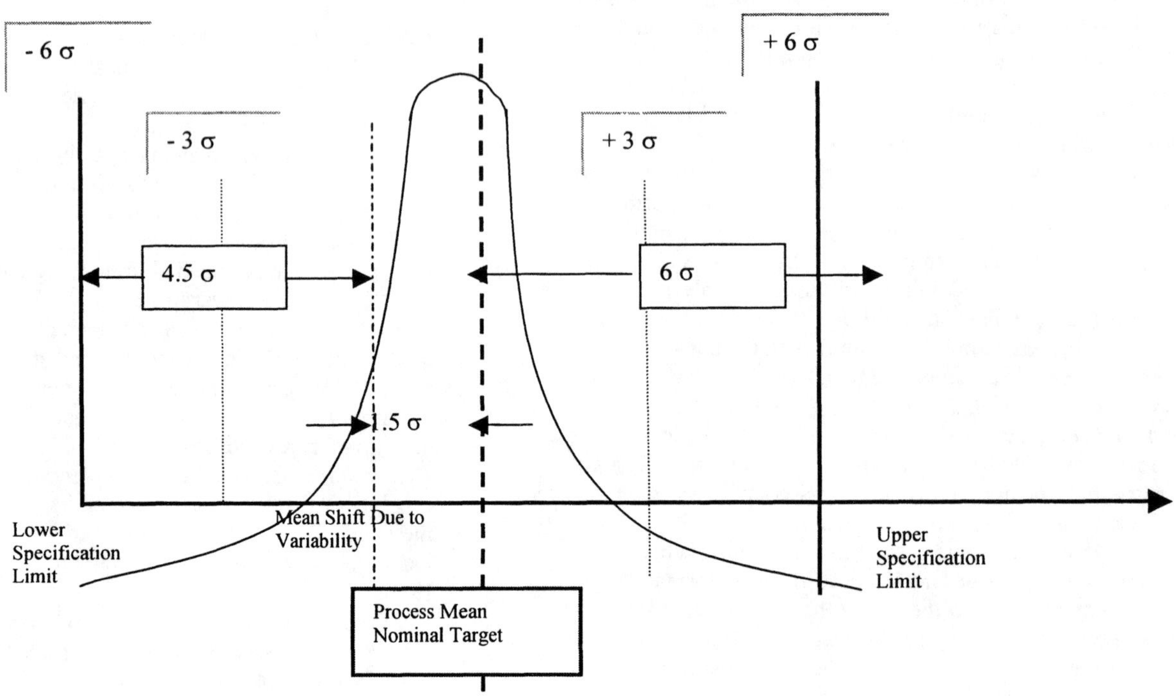

Figure 1. Process Performance

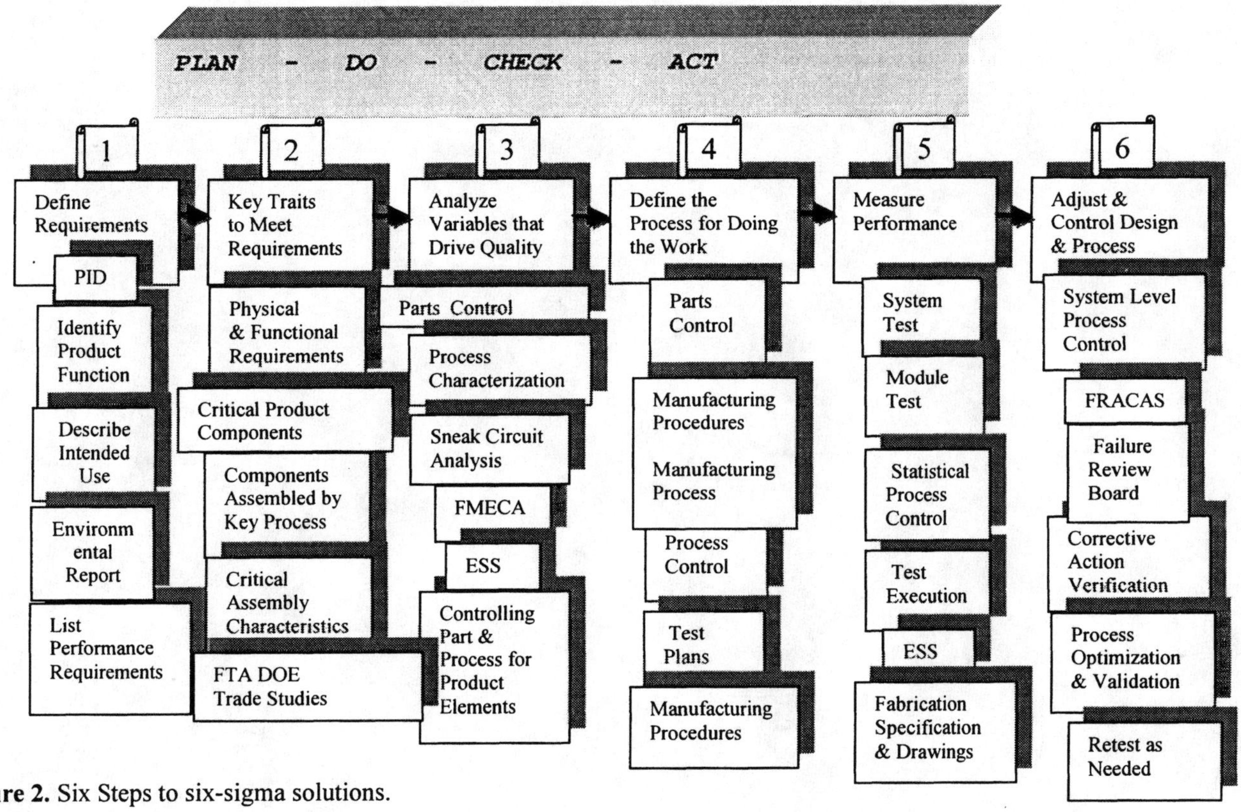

Figure 2. Six Steps to six-sigma solutions.

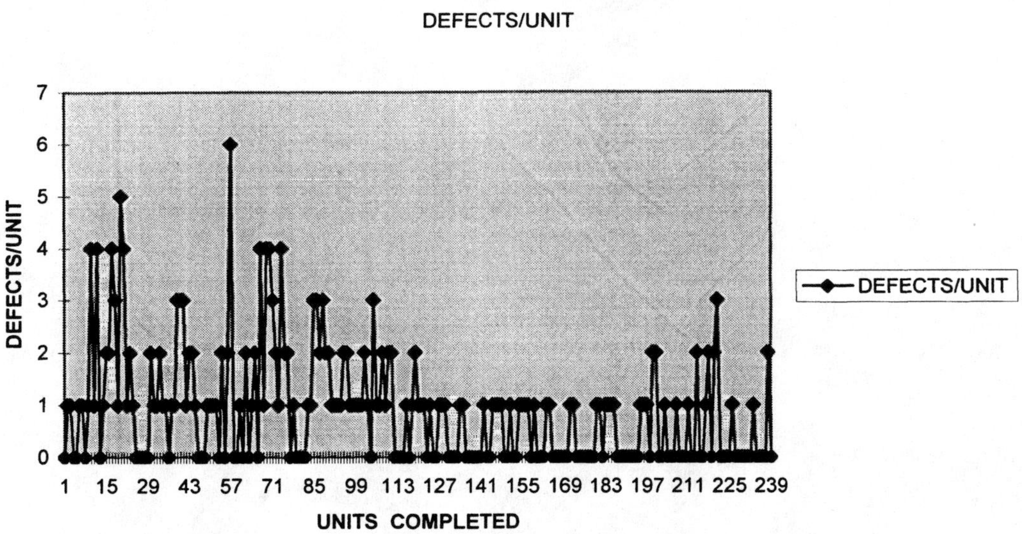

Figure 3. Chart tracking of the defective components (one sample).

Proceedings of
2001 ASME International Mechanical Engineering Congress and Exposition
November 11–16, 2001, New York, NY
SERA-Vol. 11

IMECE2001/SERA-24009

CASE STUDY - THE SAFETY OF WOOD RAILINGS

Ralph L. Barnett
Professor, Mechanical and Aerospace Engineering
Illinois Institute of Technology
Chicago, Illinois 60616

William G. Switalski
Triodyne Inc.
5950 W. Touhy Ave., Niles Illinois 60714
(847) 677-4730 Fax (847) 647-2047
(e-mail) infoserv@triodyne.com

ABSTRACT

When the handrail assembly broke away from a wooden deck attached to the rear of a private residence, the victim fell 12 feet to the lawn and sustained injuries rendering him a quadriplegic. Although the local building code required the handrail to withstand a 200 lb load applied in any direction at any point on the handrail, no guidance was given to the do-it-yourselfer who built the deck and railings to assure him that the final construction would produce an acceptable railing. The authors conducted testing and a statistical analysis of railing strength comparing the construction method used by the builder of the accident railing to another construction method utilizing a commercially available handrail bracket. The test program demonstrates that the strength of the wood used to build handrails can vary greatly and that a controlled method of building a handrail is necessary to ensure the integrity of a product intended to be consumer customized and assembled. It is necessary to have acceptable methods of railing construction because the failure of a railing joint can be life threatening. This is especially true in the consumer/do-it-yourself market where the designer/builder is not necessarily knowledgeable about building codes or construction methods. A commercially available handrail bracket known as Create-A-Rail® can provide the consumer with the guidance necessary to assure an acceptable handrail / post joint.

I. Accident Scenario

A homeowner with ordinary craft skills built a wooden deck which was elevated 12 feet above his backyard. Except for the railings, the construction features were unremarkable. The building supplier furnished railing components consisting of lathe turned spindles and posts together with a milled handrail. The project was undertaken without a building permit, a kit, standardized drawings, or advice from the building supplier. The only fasteners that were furnished were 2-1/2 inch deck screws.

After the deck was completed, an invited neighbor used it to play catch with children in the yard below. While attempting to snag a ball which had fallen short, the neighbor pushed against a corner railing as he leaned over it and reached downward; the railing broke and he fell 12 feet and became a quadriplegic. Examination of the railing system after the accident revealed that the two deck screws that secured the handrail to the posts by "toenailing" were bent but intact in the posts. Wood divots pulled from the end grain of the handrail provided an escape geometry which released the handrail.

II. Railing Construction

The railing structure involved in this case study is depicted in Fig. 1 where the toenails in the handrail were inserted into pilot holes drilled by the homeowner that angled upward and into the page:

Corner Post Screw
36° above the horizontal in the vertical plane;
13.5° from the centerline of the handrail in the horizontal plane.

Side Post Screw
26° above the horizontal in the vertical plane;
18° from the centerline of the handrail in the horizontal plane.

The handrail is 29-1/16 inches in length and supports the tops of four balusters; the bottom of the balusters engage a 2 x 4 of the same length. A single deck screw is used at each joint. The 2 x 4 is fastened into mating posts with four deck screws; two at either side. All joints in the railing system were fastened together using 2-1/2 inch deck screws and predrilled using a 1/8 inch diameter drill bit.

All components of the railing system were constructed from treated southern yellow pine. Concentrating on the handrail, its cross section has been superimposed onto a typical tree section in Fig. 2. Radical differences in grain structure are observed depending on the orientation of the lumber when sawn. Wood strength is highly dependent on the grain orientation [1].

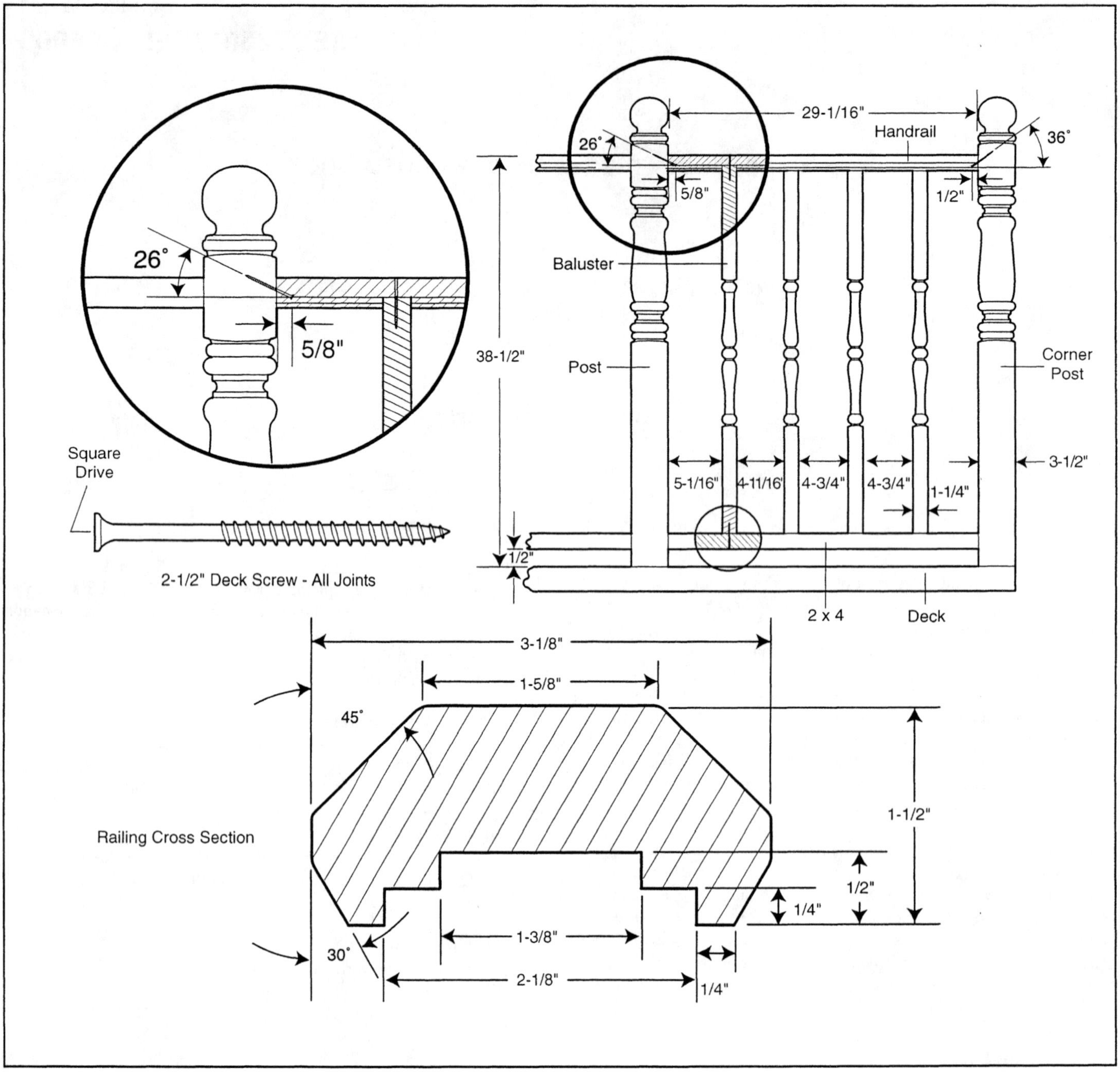

Fig. 1 - Case Study Railing System

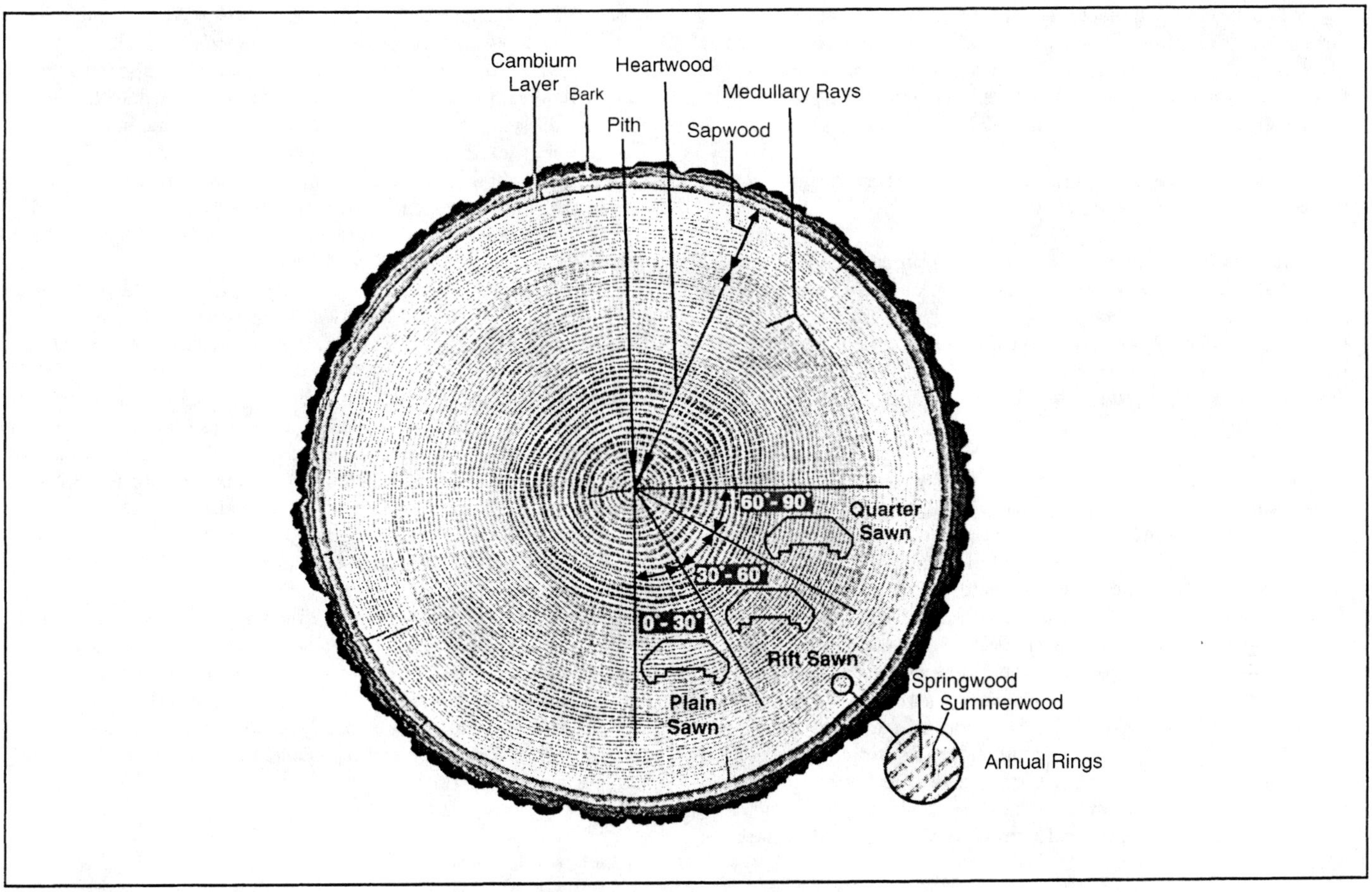

Fig. 2 - Typical Cross Section of a Tree Trunk

Tree growth takes place only in the cambium layer by cell division. It follows that an ideal tree is axisymmetric and wood is both orthotropic and nonhomogeneous. Every year long tube-like vertical fibers are formed around the entire trunk; these fibers bond to their circumferential neighbors grown in the same year and to their radial neighbors grown in the previous year. Consequently, the strength and stiffness properties are different in the axial, radial, and circumferential directions which are all perpendicular to each other. Noting that radial bonding may vary between years because of dissimilar climatic conditions, it follows that wood may not be homogeneous in the radial direction.

Because wood is obtained from many trees that are randomly sawn, and since wood is orthotropic and nonhomogeneous, the *in situ* strength of nominally identical members will vary stochastically. This statistical notion is particularly poignant when the resistance of only a small region of wood is challenged, e.g., the end of the railing where it is toenailed into a post.

III. Railing Performance Requirements – A Brief History

The first American safety standard to address railing design (except residential railings) was the *American Standard Safety Code for Floor and Wall Openings, Railings and Toe Boards*, American Standards Association ASA A12-1932 [2]. Pertinent technical terms were defined for the first time:

> ***Handrail*** *— A single bar or pipe supported on brackets from a wall or partition, as on a stairway or ramp, to furnish persons with a handhold in case of tripping.*
>
> ***Standard Railing*** *— A vertical barrier erected along exposed edges of a floor opening, wall opening, ramp, platform, or runway to prevent falls of persons.*

The standard specifically addressed wood railings, pipe railings and structural metal railings stating:

> *The anchoring of posts and framing of members for railings of all types shall be of such construction that the completed structure shall be capable of withstanding a load of at least 200 pounds applied in any direction at any point of the top rail. [Section 7.3(d)]*

This requirement remained unchanged in the 1967 revision of the safety standard, United States of America Standard USAS A12.1-1967

[3]. When the Occupational Safety and Health Act became law in 1971, the 1967 revision of the safety standard was adopted and remains the law in effect today. However, confusion sometimes arises because the federal government never updated its regulations to keep pace with the subsequent revisions of A12.1 in 1973 and its successor standard, A1264.1 in 1989 and 1995.

In the 1973 revision of A12.1 [4], the railing performance requirement was completely revamped to read:

The completed railing shall be designed and constructed to withstand a load of 25 pounds per lineal foot applied in any direction at the top railing. The intermediate rail shall be capable of withstanding a horizontal load of 20 pounds per lineal foot. The end of terminal posts shall be capable of withstanding a load of 200 pounds applied in any direction at the top of the post. The above loads are not additive. [Section 7.3.1]

New terminology also appeared in 1973 when "Guard Railing" replaced the previously used "Standard Railing:"

The Guard Railing is an assembly consisting of a top rail, intermediate rail and posts with vertical height in the range of 36 to 42 inches.

Concurrent with the A12.1 standard for railings was the USAS A64.1-1968 [5] standard for Fixed Industrial Stairways. The A12.1 and A64.1 code committees combined in 1989 to form the A1264.1 committee and published a revised standard entitled, *American National Standard Safety Requirements for Workplace Floor and Wall Openings, Stairs, and Railing Systems* [6]. The scope of the new standard was limited to industrial and workplace situations. Construction, residential and commercial occupancies are excluded except where necessary maintenance or workstation access may be required.

Once again, new technical terminology emerged to replace the previous "Guard Railing:"

Guardrail/Railing System/Stair Railing System *— Framework of vertical, horizontal or inclined members, grillwork or panels, or combinations thereof, supporting a handrail and acting as a safety barrier for protection of persons at or near the outer edge of stair, ramp, landing, platform, hatchway, manhole, or floor opening.*

Handrail *— Horizontal, sloping, or vertical member normally grasped by hand for support. This member may be part of a railing system and is often, but need not be, a top member (top rail) of a railing system.*

The Railing System is an assembly consisting of a top rail, intermediate rail or equivalent protection, and posts having a vertical height in the range of 40 to 44 inches (another significant revision). In addition, the strength requirements for a Railing System were changed from a load based criteria to a deflection based criteria:

The completed railing systems shall be designed and constructed for its intended use to preclude system failure. As a minimum, it shall withstand a concentrated load of 200 pounds (90.7 kg) applied in any direction, except upward, at the mid-point between posts without exceeding maximum allowable deflection. The intermediate rail shall be capable of withstanding a horizontal load of 80% of the above stated load applied at mid-point and mid-height without exceeding the maximum allowable deflection. The end of terminal post shall be capable of withstanding a load of 200 pounds (90.7 kg) applied in any direction at the top of the post. The above loads are not additive. [Section 5.6.1]

The 200 pound load specified in the standard is applied at the mid-point of the railing span because this is the location where the deflection in the rail will be the greatest. Although the maximum allowable deflection is left to the Railing System designer, the standard offers the suggestion (not a requirement) that a residual deflection in excess of one-half inch may indicate potential failure. [Ref. Section E5.6.1]

The Railing System requirements remained unchanged in the 1995 revision of the A1264.1 safety standard [7].

It should be observed that American National Standards Institute safety standard ANSI A1264.1-1995 is not as stringent as its predecessor ASA A12 -1932 or the current Occupational Safety and Health (OSHA) Standard 29 CFR Ch.XVII.(7-1-99 Edition) 1926.451(g)(4)(vii) [8]. Two hundred pounds applied midspan gives rise to an end shear of only 100 lbs in the handrail. On the other hand, when 200 lbs is applied at every point along the handrail, the critical end shear is developed when the load is located next to the post in a horizontal direction; here the shear is 200 lbs or double that specified by the current ANSI standard. At present, the highest standard for handrail design is set by the Building Officials and Code Administration BOCA National Building Code, 1993 [9]:

1615.8.1 Handrail design and construction*: Handrails shall be designed and constructed for a concentrated load of 200 pounds (91 kg) applied at any point and in any direction. Handrails located in other than dwelling units in occupancies in Use Groups R-2 and R-3 shall also be designed and constructed for a uniform load of 50 pounds per foot (74 kg/m) applied in any direction. The concentrated and uniform loading conditions shall not be applied simultaneously.*

IV. Toenail Joint

When an outward facing horizontal concentrated load is applied to the right end of the handrail shown in Fig. 1, it is resisted entirely by the toenailed deck screw at the right post; none of the other thirteen joints in the railing system feel any load at all. Since this is the "worst case" loading of the railing system, the strength of the toenail joint is equivalent to the strength of the entire railing structure. This strength was characterized with five exemplary railings that were constructed to be nominally identical to the artifact. Southern Yellow Pine railings were utilized during testing to be consistent with the accident railing. These were located in the corners of the 8 ft x 8 ft test deck shown in Fig. 3. Using the test setup depicted in Fig. 3, the handrails were tested to failure using a horizontal force applied 3 inches from the corner post. The force was applied tangential to the top handrail surface through a manual winch. The magnitude of the applied force was measured with a Chatillon digital force gauge and accompanying load cell.

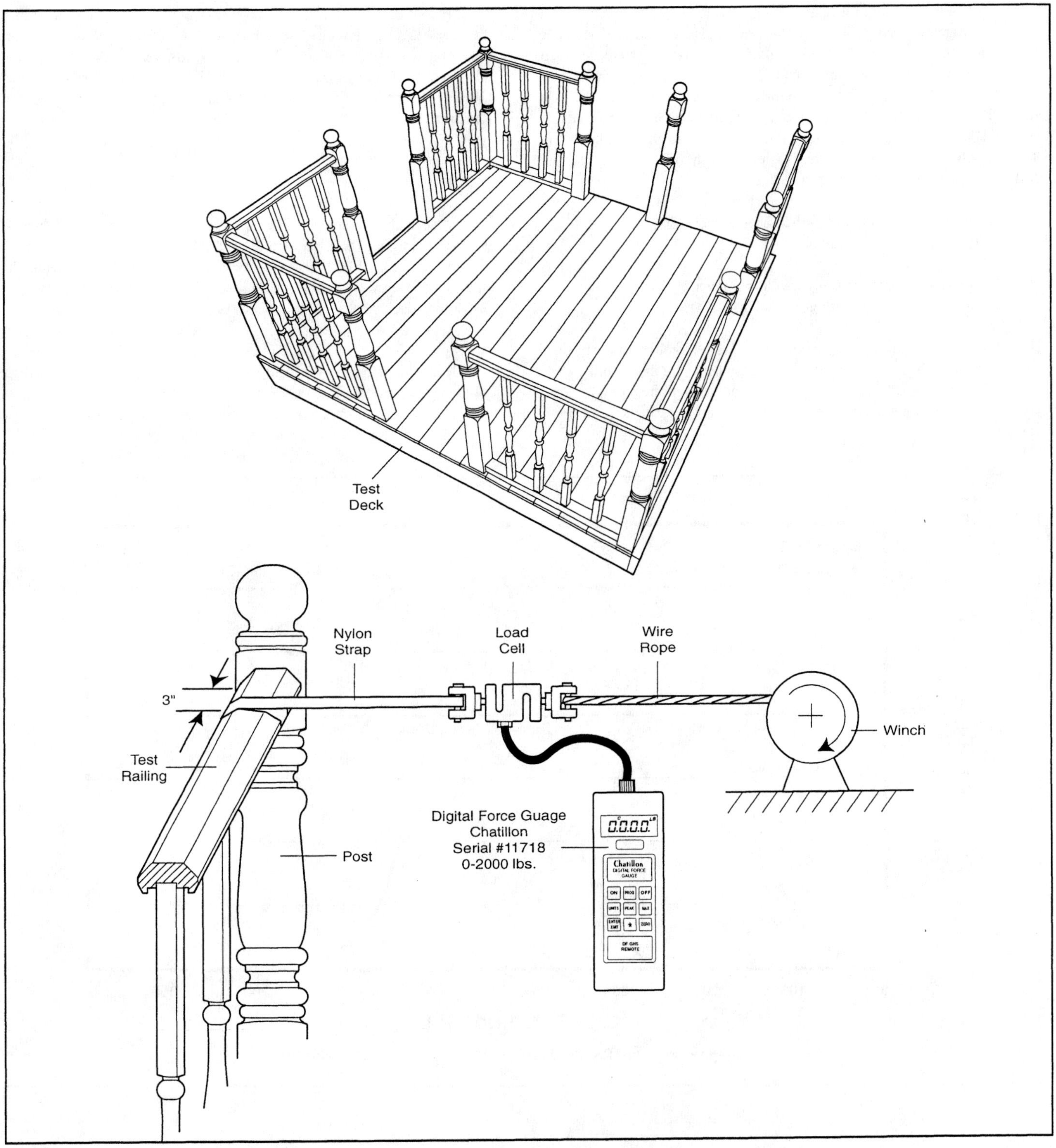

Fig. 3 - Test Deck and Instrumentation

The five fracture loads measured for the five toenail joints are recorded in Table I together with their mean, standard deviation, and coefficient of variation. These data lead to a "bell shaped" curve that may be represented by a Gaussian Distribution Function [10] as illustrated in Fig. 4. The area beneath the probability density curve from the extreme left up to 200 lbs is the probability of failure under a load of 200 lbs or the percentage of railings that will break under a 200 lb load; the same applies to any load including 100 lbs which is important here. Using conventional procedures, the following statistical inferences can be drawn:

a. The probability of the railing fracturing under a 200 lb load is 26.3%, i.e., about one chance in four.

b. The probability of the railing fracturing under a 100 lb load is 2.43%, i.e., about one chance in forty.

A 200 pound man wearing work shoes cannot develop a static push of 100 lbs on a railing. Consequently, a homeowner or an inspector has less than one chance in forty of ever detecting a bad railing joint under extreme exertion. On the other hand, even mild dynamic impacts on the railing will develop 100 lbs.

In each of the five tests, the end grain was pulled from the handrail and the deck screws remained in the posts; in all cases, their appearance resembled the artifact. The higher failure loads were associated with more closely spaced growth rings in the handrail lumber. As a final observation, the 30% coefficient of variation is quite high and indicates very large variability in the fracture loads.

V. Metal Bracket

A specialized metal bracket manufactured under the name Create-A-Rail® is available for fastening the handrail to the posts. This connector is depicted in Fig. 5. The bracket is screwed directly into the post using three similar deck screws (5a). The handrail sits in a saddle and is secured with two 2-1/2 inch deck screws toenailed into both sides (5b).

Five railing systems were constructed with the same wooden components shown in Fig. 1. They were strength tested using the same test deck and testing protocol adopted for the "toenail" joint. The results of these tests are tabulated in Table I; the associated Gaussian or normal probability density curve is shown in Fig. 4. The following statistical inferences may be drawn from the normal distribution function:

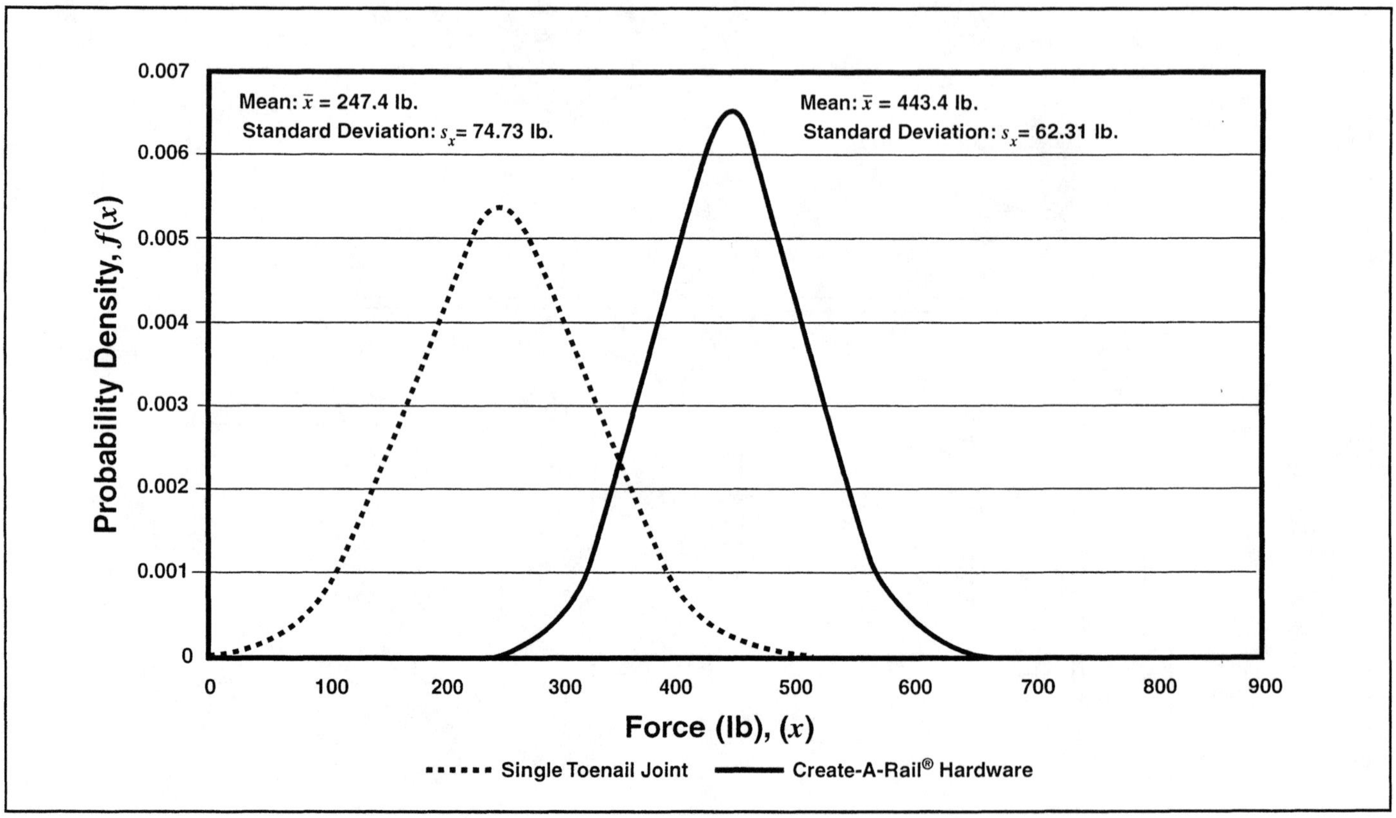

Figure 4 - Gaussian Strength Distribution: $f(x) = \frac{1}{s_x\sqrt{2\pi}} e^{-\frac{1}{2}\left(\frac{x-\bar{x}}{s_x}\right)^2}$

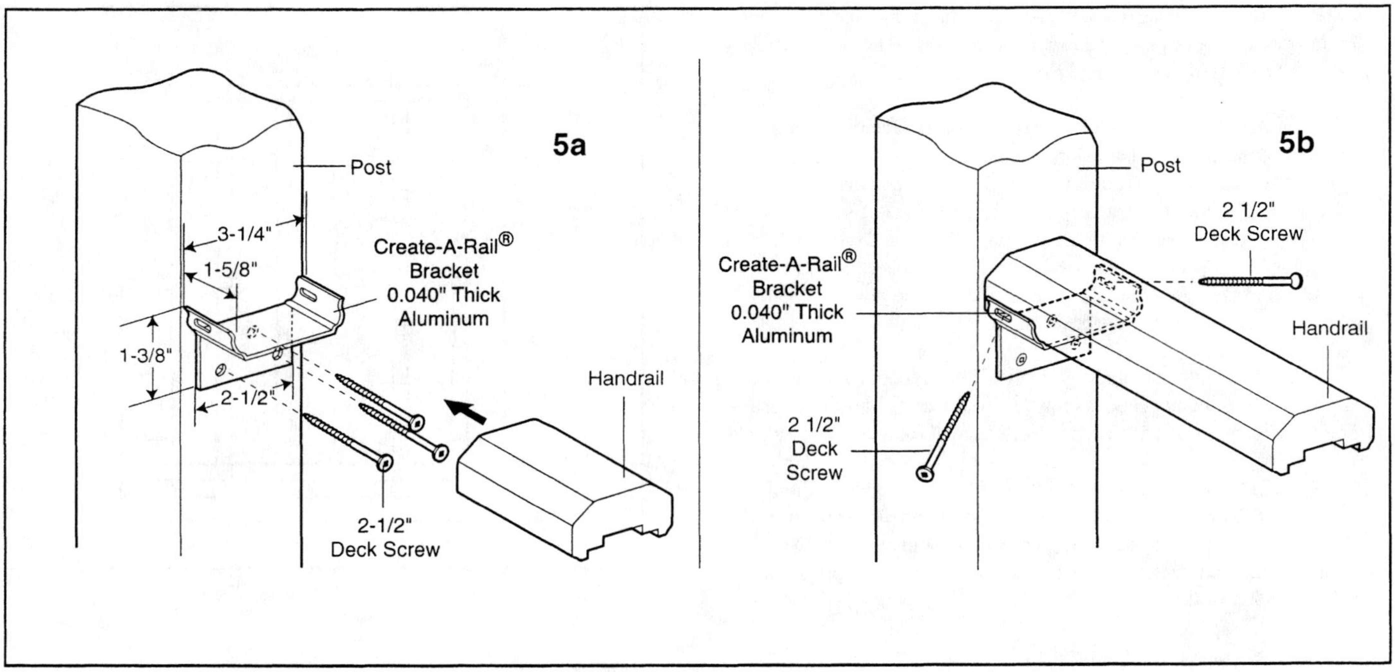

Fig. 5 - Create-A-Rail ® Connector

a. The probability of the railing system fracturing under a 200 lb load is 4.6884 x 10^{-5}, i.e., one chance in 21,329.

b. The probability of the railing system fracturing under a 100 lb. load is 1.7869 x 10^{-8}, i.e., one chance in 55,962,841.

Table I. Railing Strength, lb

Handrail Specimen	Toenail Joint	Metal Connection
1	250	
2	297	
3	215	
4	334	
5	141	
6		417
7		351
8		487
9		509
10		453
Mean Fracture Load, $\bar{x}$	247.40 lb	443.40 lb
Standard Deviation, s_x	74.73 lb	62.31 lb
Coefficient of Variation, $s_x/\bar{x}$	30.2%	14.1%

Comparing the Create-A-Rail® connector to the "toenail" joint shows that under a 200 lb load, the Create-A-Rail® will fail once in every 21,000 applications whereas the "toenail" joint fails in one out of four railings. For 100 lbs, the Create-A-Rail® virtually never fails; however, the "toenail" joint fails on the average once in every forty railings. The Create-A-Rail® design allows four times the purchase of the deck screws compared to the "toenail" joint; it's much more difficult to encounter a "weak" spot. As shown in Fig. 5b, the saddle alone provides lateral resistance independent of the wood properties.

The stronger railings associated with the Create-A-Rail® connectors once again displayed closely spaced growth rings. The scatter in fracture load data, as measured by the 14% coefficient of variation, is much smaller than that associated with the "toenail" joint.

VI. Discussion and Conclusions

1. Handrails and railing systems should be designed and constructed for a concentrated load of 200 pounds applied at any point and in any direction on the top rail. This traditional requirement addresses the railing / post joint more stringently than modern railing standards.

2. One of the usual top rail fabrication systems for the subject railing involves the use of six screws; four into each of the four balusters and a toenail screw into each of two vertical posts. Six 2-1/2 inch deck screws in such a short railing span will generally be conceived of as a very substantial construction by a homeowner or do-it-yourself craftsman.

3. Forces applied to the "toenailed" railing in the upward or downward direction are resisted by an extremely efficient structural system. The

standard 200 pound loading will always be accommodated in the vertical plane. Checking the railing by pushing down or lifting up on the handrail will always produce a tactile feedback of a rigid member.

4. If a concentrated horizontal force is applied at the center of the top rail, each post connection resists the load equally due to symmetry. If the load is applied at either end of the railing, all of the load is transferred to a single post and a single joint adjacent to the load. This latter result is true regardless of the type of end connection (e.g. simply supported or fixed-fixed). Note that railing design follows the "worst case scenario" philosophy; a 200 pound load in any direction at any location along the railing. Clearly, the horizontal force applied near the post is the critical railing load.

5. Differences among nominally identical trees, random trunk location of sawn lumber, nonhomogeneous growth patterns, and orthotropic behavior beget the exceptional variability in wood strength found in small samples.

6. The single "toenail" joint between the post and handrail is literally the weakest link " in the railing. When this fact is combined with the stochastic nature of wood, the expected railing performance will exhibit high variability. The railing addressed in this case study violates the 200 pound load standard 26.3% of the time.

7. The "toenail joint will fail under a 100 pound load 2.43% of the time, i.e., once in every 20 railings (two joints each). A strength level in the neighborhood of 100 pounds is virtually undetectable. Neither a home craftsman nor a building inspector can challenge this resistance without special loading.

8. The Create-A-Rail® connector will almost never fail under a load of 100 pounds (one chance in 56 million). Only one in 21,000 connectors will fail below 200 pounds. This superior performance is reflected in both higher strength and small variability. The forgiving nature of the Create-A-Rail® design is attributable to the saddle construction together with five deck screws that are optimally deployed. The Create-A-Rail® bracket design helps to compensate for errors in railing construction by indicating the use of 5 screws and adding a metal component to the railing / post connection.

9. Effective alternate designs can be fashioned without specialized hardware. For example, a six baluster design, Fig. 6, may be employed which places a baluster flush against each post. Two or three deck screws may be driven through this member directly into the post. In the normal manner, one or two vertical deck screws will secure the handrail to the baluster which fits into the milled slot on the bottom of the handrail. This provides an interference against any outward movement.

10. The handrail/post connection in a wooden railing system is a critical structural joint whose failure is life threatening. Do-it-yourself homeowners will not understand how to control the "weakest link" and stochastic nature of this connection. Furthermore, they will not be able to properly proof test the joint after construction to assure its structural integrity. Safety demands that proper instructions and warnings accompany the sale of the railing components.

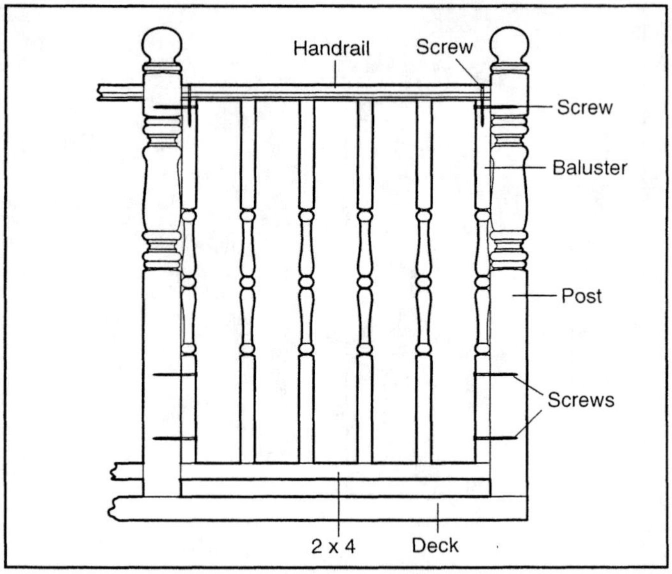

Fig. 6 - Six Baluster Design

REFERENCES

[1] Architectural Woodwork Quality Standards; 7th edition version 1.0, Architectural Woodwork Institute, Isaac Newton Square, Reston, Virginia 20190; 1997.

[2] "American Standard Safety Code for Floor and Wall Openings, Railings, and Toe Boards," *ASA A12 -1932*. New York: American Standards Association, approved May 3, 1932.

[3] "USA Standard Safety Requirements for Floor and Wall Openings, Railings, and Toe Boards," *USAS A12.1-1967*. New York: American National Standard Institute, approved March 17, 1967.

[4] "American National Standard Safety Requirements for Floor and Wall Openings, Railings, and Toeboards," *ANSI A12.1-1973*. New York: American National Standards Institute, approved March 15, 1973.

[5] "Requirements for Fixed Industrial Stairs," *USAS A64.1-1968*. New York: American National Standards Institute, approved February 16, 1968.

[6] "American National Standard Safety Requirements for Workplace Floor and Wall Openings, Stairs, and Railing Systems", *ANSI A1264.1-1989*. New York: American National Standards Institute, approved October 5, 1989.

[7] "Safety Requirements for Workplace Floor and Wall Openings, Stairs and Railing Systems," *ANSI A1264.1-1995*. New York: American National Standards Institute, approved February 6, 1995.

[8] "Fall Protection," 29 CFR 1926.451(g)(4)(vii). Washington, D.C., Occupational Safety and Health Administration; Chapter XVII, 7/1/99 edition.

[9] "Special Loads," Section 1615.0, *The BOCA National Building Code/ 1993*. Country Club Hills, IL: Building Officials & Code Administration International, Inc. 1993, pp. 185 – 186.

[10] Devore, Jay L., Probability and Statistics for Engineering and the Sciences; Chapter 4.2, Cumulative Distribution Functions and Expected Values, and Chapter 4.3, The Normal Distribution, pp. 150 – 171.

Proceedings of
2001 ASME International Mechanical Engineering Congress and Exposition
November 11–16, 2001, New York, NY
SERA-Vol. 11

IMECE2001/SERA-24010

THE IMPACT OF SAFETY MITIGATION TO RELIABILITY OF CO-GENERATION UNITS

Joseph P. Balkey
1330 Wilkie Drive
Charleston West Virginia
304/345-1699
JPBASB@aol.com

ABSTRACT

Companies may be considering improving their power reliability by installing one or more new co-generating unit(s) in addition to buying power. However, they may be unaware that the safety devices that protect the unit(s) may also inadvertently shutdown the unit(s). This paper first describes how a co-generation unit operates and some of its safety features. One of the safety features is to shutdown the unit when a temperature is inadvertently high. However, the number of instruments used to measure a temperature greatly impacts both safety and reliability. This paper graphically shows the simultaneous impact of four types of voting to both safety and reliability and discusses the purpose of each one. Knowing the purpose of each type of voting can lead to selecting a co-generation unit that matches a users needs. This paper also shows how the values in the graph are calculated.

INTRODUCTION

A few years ago, a facility with many large stand-alone production units bought two off-the-shelf co-generation machines. Each machine burns natural gas to supply electricity and steam to the many large stand-alone production units. In most circumstances, one co-generation machine can be out-of-service without impacting any of the production units; but if both machines are out-of-service, most of the production units will be impacted. There was no steam backup and a public utility could supply some backup electricity in the winter but not in the summer.

The availability of each co-generation machine approached its 96% design objective, but the probability that at least one of the two machines will be running was less than desired. As a result, I was asked to assess potential projects that will improve the probability that at least one machine will always be running to produce steam and electrical power for the plant. During this study, the relationship between safety and reliability was identified as a design planning error.

PROCESS DESCRIPTION

A co-generation machine burns natural gas to produce electricity and steam in two distinct processes; an auxiliary system and a gas turbine-generator system. The auxiliary system produces steam and the gas turbine-generator produces electrical power.

Auxiliary system

This system produces (high, medium, and low pressure) steam by burning natural gas to heat and vaporize (high, medium, and low pressure) boiler feedwater in a Heat Recovery Steam Generator. The three steam pressures are distributed throughout the plant. (The high pressure steam header can generate some electrical power by dropping high pressure steam to a lower pressure through a steam turbine).

One deaerator removes oxygen from the boiler feedwater before it feeds a Heat Recovery Steam Generator at three locations. Three pumps in parallel feed boiler feedwater at high and medium pressures to make high and medium pressure steam. Two pumps in parallel feed boiler feedwater at low pressure to make low pressure steam.

Lube oil, seal flush, and cooling support operating the pumps. Utilities (instrument air, cooling water, and blowdown water) operate valves and other components.

Gas turbine-generator system

This system produces electricity in a generator by rotating windings using a turbine. The hot gases exhausted from burning the natural gas are compressed, further heated by combusting more natural gas, and then dropped in pressure through a turbine.

This system also has utilities (lube oil, air, fuel gas, hydraulics, instrument are) that support it.

SAFETY SYSTEMS

The safety systems differ for each system and it's the safety system for the Gas turbine-generator which is the topic of this paper.

Auxiliary system safety system

If this system has an operating upset, it has controls to prevent damaging equipment (**wanted shutdown**). If these controls fail, an upset could damage the system (**unwanted accident**). However, when these controls fail, they do not inadvertently shutdown the operation (**unwanted false shutdown**).

> For example, if the liquid level on the suction of the high pressure boiler feedwater pumps inadvertently drops, an instrument measuring the level should stop the pumps to prevent damaging the pumps (**wanted shutdown**). If the level measuring instrument fails and then the level decreases, the pumps will be damaged (**unwanted accident**). However, a false liquid level measurement will inadvertently shutdown (trip) the pumps (**unwanted false shutdown**).

Gas turbine-generator system safety

If this system has an upset, it has controls to prevent damaging equipment (**wanted shutdown**). If these controls fail, an upset could damage the system (**unwanted accident**). However, when some of these controls fail, they inadvertently shutdown the operation (**unwanted false shutdown**). Unwanted shutdowns of this system are the subject of this paper.

> As one example, if a flashback increases the temperature in the combustor, the thermocouple measuring the temperature should unload the turbine to prevent damaging it (**wanted shutdown**). If the thermocouple fails and then the temperature rises, the turbine will be damaged (**unwanted accident**). However, a false temperature measurement will inadvertently unload the turbine (**unwanted false shutdown**).

> As an another example, if there is a fire in one of four areas of the Combustion Turbine, the thermocouple measuring the temperature should shutdown (trip) the turbine to prevent further damage (**wanted shutdown**). If the thermocouple fails and a fire increases the temperature, the Combustion Turbine will be damaged (**unwanted accident**). However, a false temperature measurement will inadvertently trip the Combustion Turbine (**unwanted false shutdown**).

Voting

A temperature can be measured using one, two, or three instruments to shutdown the system .

- when one instrument measures a temperature and trips the unit if it senses a high temperature, this is 1-oo-1 voting.

- when two instruments measure a temperature and trip the unit if either one of them sense a high temperature, this is 1-oo-2 voting.

- when two instruments measure a temperature and trip a unit only if both of them sense a high temperature, this is 2-oo-2 voting.

- when three instruments measure a temperature and trip a unit if any two of them sense a high temperature, this is 2-oo-3 voting.

The auxiliary system has less than five instruments primarily designed with either 1-oo-1 voting or 1-oo-2 voting but the gas turbine/generator has about 50 temperature instruments primarily designed with either 1-oo-1 voting or 1-oo-2 voting.

TECHNICAL TOOL (voting impact on **unwanted** shutdowns)

Figure 1 compares the frequency of **unwanted false shutdowns** (reliability), **unwanted accidents** (safety), and **wanted shutdowns** across four voting configurations (1-oo-1, 1-oo-2, 2-oo-2, 2-oo-3). The shutdown frequencies that are plotted in Figure 1 are numerated in Table 1 and calculated from fault trees shown in the **Appendix**.

> Comparing the frequencies of **unwanted false shutdowns** and **unwanted accidents** for each voting configuration explains why the two co-generation units (designed with 1-oo-2 voting) experience frequent **unwanted false shutdowns**. 1-oo-2 voting produces the least frequent **unwanted accidents** but the most frequent **unwanted false shutdowns**.

> Comparing the frequencies also shows that to improve the situation, it is better to convert to 2-oo-3 voting rather than to 1-oo-1 or 2-oo-2 voting. 2-oo-3 voting reduces the frequency of **unwanted false shutdowns** without sacrificing **unwanted accidents**. Whereas 1-oo-1 voting is the worst compromise of **unwanted false shutdowns** (reliability) and **unwanted accidents** (safety) and 2-oo-2 voting has the least frequent **unwanted false shutdowns** but the most frequent **unwanted accidents**.

CALCULATIONS

This section describes how the values in the fault trees shown in the Appendix are calculated. For comparison purposes, the system is assumed to have an upset 1/yr. The 0.59/yr failure rate of a temperature sensor is from a database and the test interval of a sensor is assumed to be 2 months. Because the sensor could fail anytime between the tests, it is assumed to fail halfway between a 2 month test interval.

Shutdown scenarios (basis)

Given a system that has a temperature sensor(s) to shut it down if its temperature rises can configure those sensors to operate at least one of four ways. However, no matter which configuration is selected, in addition to the **wanted** shutdowns, the system can be shutdown from **unwanted** false signals (reliability) or the system can be shutdown by an **unwanted accident** if the sensors fail to shutdown the system (safety).

Wanted shutdowns (desired operation). Assumes the system temperature inadvertently rises 1/yr and one or more temperature sensors shutdown the system.

Unwanted false shutdowns (reliability). Assumes the system is operating at its normal temperature when a temperature sensor **fails high** 0.59/yr and **inadvertently shutsdown** the system. This reduces system reliability.

Unwanted accidents (safety). Assumes the system temperature inadvertently rises 1/yr and one or more temperature sensors have a 5% chance of **failing low** and the system **fails to shutdown.** The chance of failing low assumes that a sensor is tested every 2 months.

$$0.05 = 0.59/yr \times (2\ mo/2 \times yr/12\ mo)$$

Voting configurations impact on unwanted shutdowns

The frequencies of **unwanted false shutdowns** (reliability) and **unwanted accidents** (safety) are described and calculated for four voting configurations. For each configuration, the system is first described as it is intended to **operate** and then the **unwanted false shutdown** and the **unwanted accident** are described with the calculated frequencies. These frequencies help to compare designs. Although the absolute values will vary widely for individual designs, the relative comparisons should remain unchanged.

Voting 1-oo-1 configuration. A system has one temperature sensor and it shutsdown the system with 1-oo-1 voting if its temperature rises.

When the system operates at normal temperature, the fault tree in **Figure A-1A** calculates that if the one temperature sensor fails high, the sensor will inadvertently shutdown the system 5.9e-1/yr.

If the system temperature inadvertently rises 1/yr, the fault tree in **Figure A-1B** calculates that if the one temperature sensor fails low, the system will fail to shutdown 5.0e-2/yr.

Voting 1-oo-2 configuration. A system has two temperature sensors and either shutsdown the system with 1-oo-2 voting if its temperature rises.

When the system operates at normal temperature, the fault tree in **Figure A-2A** calculates that if either of the two temperature sensors fails high, one of two sensors will inadvertently shutdown the system 1.18e-0/yr.

If the system temperature inadvertently rises 1/yr, the fault tree in **Figure A-2B** calculates that if both temperature sensors fail low, the system will fail to shutdown 2.5e-3/yr.

Voting 2-oo-2 configuration. A system has two temperature sensors and both are needed with 2-oo-2 voting to shutdown the system if its temperature rises.

When the system operates at normal temperature, the fault tree in **Figure A-3A** calculates that if one temperature sensor had failed high when the other one fails high, two combinations will inadvertently shutdown the system 5.9e-2/yr.

If the system temperature inadvertently rises 1/yr, the fault tree in **Figure A-3B** calculates that if either of the two temperature sensors fail low, the system will fail to shutdown 1.0e-1/yr.

Voting 2-oo-3 configuration. A system has three temperature sensors and any two are needed with 2-oo-3 voting to shutdown the system if its temperature rises.

When the system operates at normal temperature, the fault tree in **Figure A-4A** calculates that if one of two temperature sensors had failed high when the third one fails high, three combinations will inadvertently shutdown the system 1.77e-1/yr.

If the system temperature inadvertently rises 1/yr, the fault tree in **Figure A-4B** calculates that if any one of three combinations of two temperature sensors fail low, the system will fail to shutdown 7.5e-3/yr.

CONCLUSION

Before buying a co-generation unit, know the voting configuration in order to install the one that fits your needs. When the facility bought the two off-the-shelf co-generation units to support the many large stand-alone production units, there was a design planning error. Off-the-shelf co-generation units are usually bought by large utility companies that have three or more co-generation units. Because they have multiple co-generation units, if an instrument inadvertently creates an unwanted false shutdown, the remaining units can continue to supply steam and power. Therefore, 1-oo-2 voting is wanted to best protect the co-generation unit. However, when only one or two co-generation units supply most of the steam and power to many large production units, 2-oo-3 voting configuration is wanted to keep the production units running.

FIGURE 1: Voting's simultaneous impact on a system's safety and reliability

Installed co-generation units were designed with 1-oo-2 voting

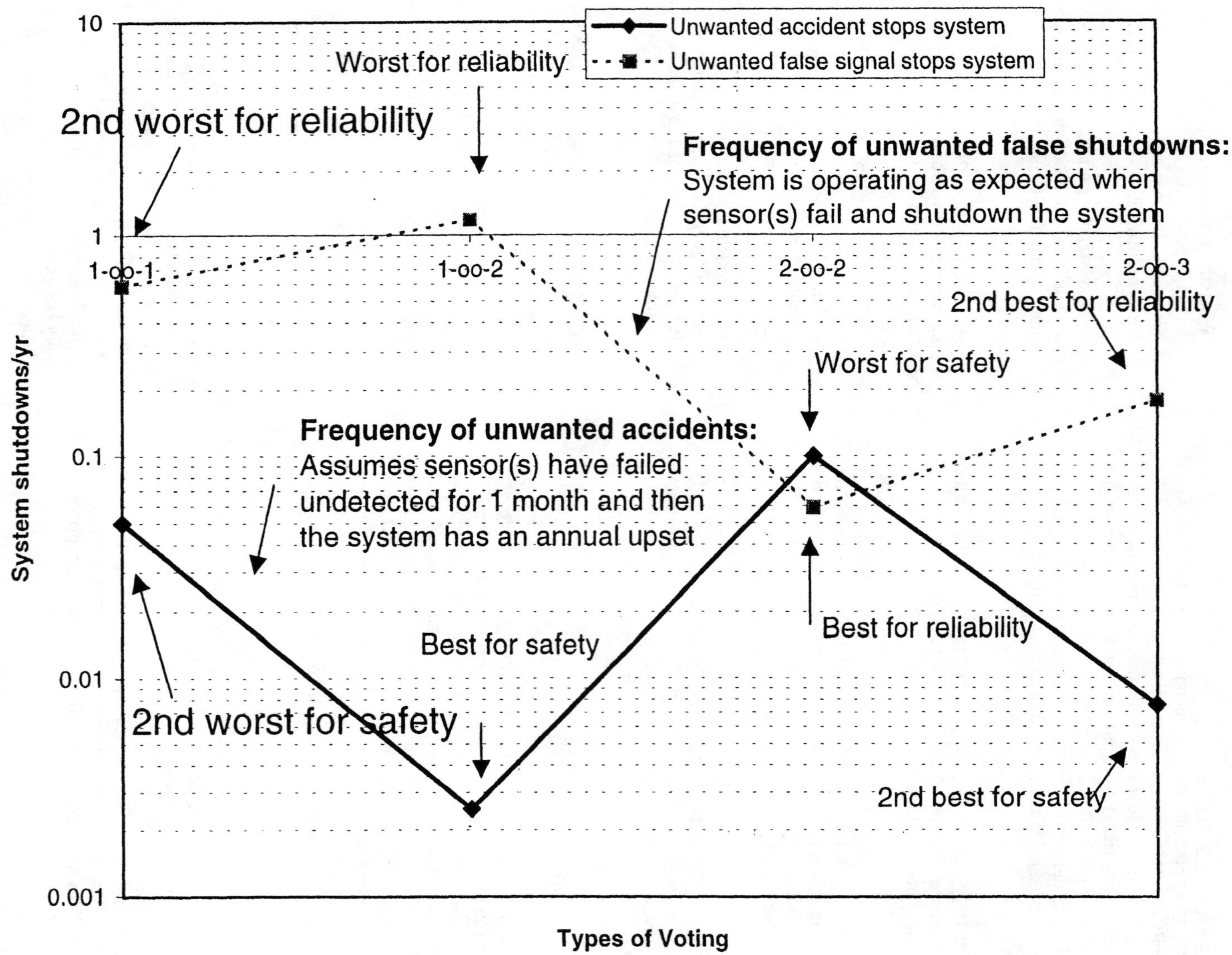

TABLE 1: SUMMARY OF VOTING CONFIGURATIONS IMPACT ON SHUTDOWN FREQUENCIES

Voting Configurations	**1-oo-1** (**worst** compromise of reliability and safety)	**1-oo-2**	**2-oo-2**	**2-oo-3** (**best** compromise of reliability and safety)
Frequency of **unwanted false shutdowns** (reliability)	0.59/yr (Fault tree 1A) **2nd worst for reliability**	1.18/yr (Fault tree 2A) **worst for reliability**	0.059/yr (Fault tree 3A) **best for reliability**	0.177/yr (Fault tree 4A) **2nd best for reliability**
Frequency of **unwanted accidents** (safety)	0.05/yr (Fault tree 1B) **2nd worst for safety**	0.0025/yr (Fault tree 2B) **best for safety**	0.1/yr (Fault tree 3B) **worst for safety**	0.0075/yr (Fault tree 4B) **2nd best for safety**
Frequency of **wanted shutdowns** (sensors work)	1.0/yr	1.0/yr	1.0/yr	1.0/yr

APPENDIX

This Appendix shows the fault trees that calculate the frequencies stated in **Figure 1** and **Table 1.**

Figure A-1A: An unwanted shutdown with 1 out of 1 voting

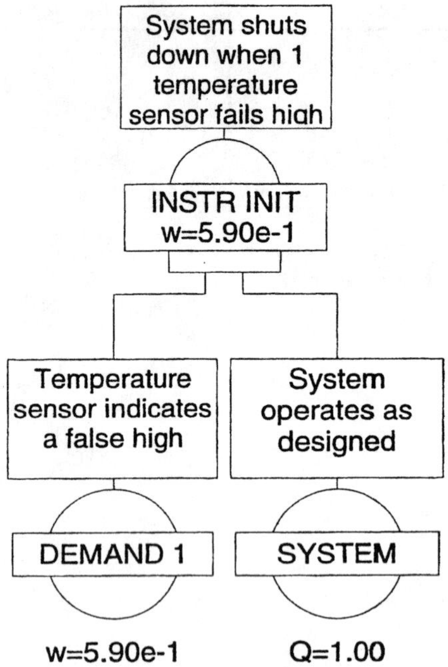

w - is a frequency value (failures/yr)
Q - is a probability value

Figure A-2A: An unwanted shutdown with 1 out of 2 voting

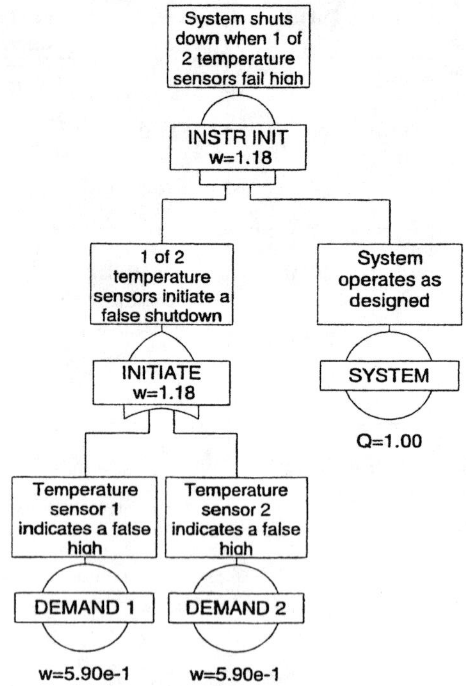

w - is a frequency value (failures/yr)
Q - is a probability value

Figure A-3A: An unwanted shutdown with 2 out of 2 voting

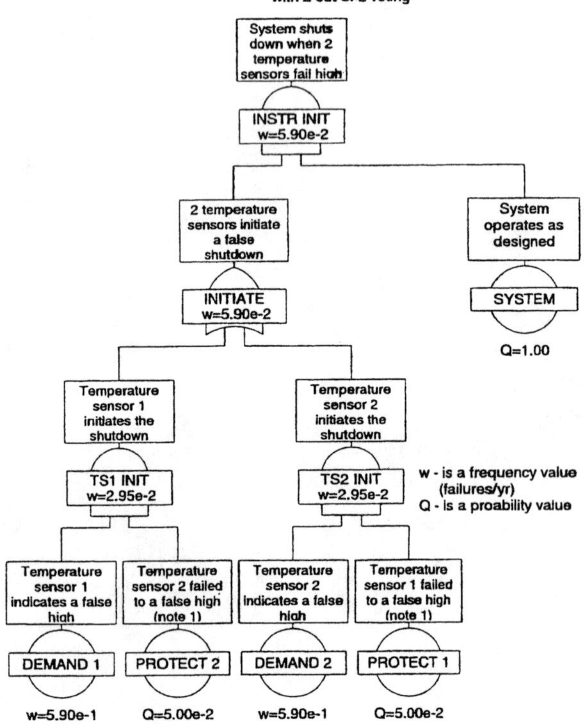

NOTES:
1. 5.0 E-2 = 0.59/yr x (2 mo/2 x yr/12 mo) test interval = 2 mo

Figure A-4A: An unwanted shutdown with 2 out of 3 voting

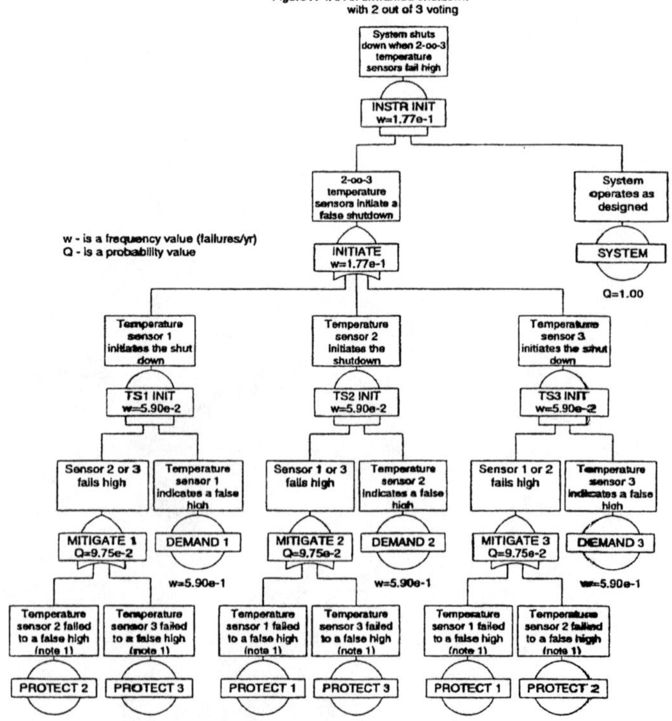

NOTES:
1. 5.0 e-2 = 0.59/yr x (2 mo/2 x yr/12 mo) test interval = 2 mo

Figure A-1B: An unwanted accident with 1 out of 1 voting

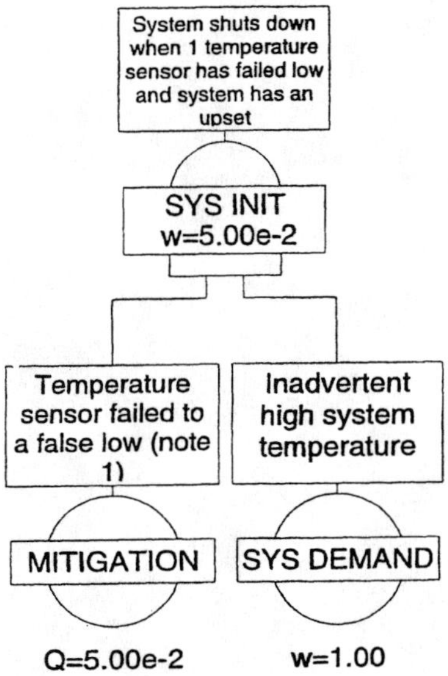

w - is a frequency value (upsets/yr)
Q - is a probability value

NOTES:
1. 5.0 e-2 = 0.59/yr x (2 mo/2 x yr/12 mo)
test interval = 2 mo

Figure A-2B: An unwanted accident with 1 out of 2 voting

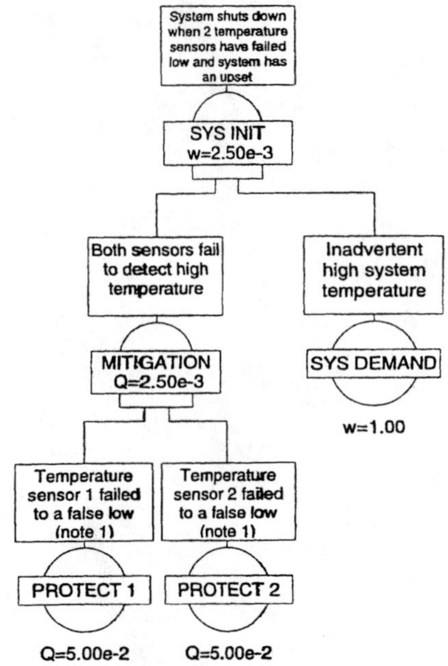

w - is a frequency value (upsets/yr)
Q - is a probability value

NOTES:
1. 5.0 e-2 = 0.59/yr x (2 mo/2 x yr/12 mo)
test interval = 2 mo

Figure A-3B: An unwanted accident with 2 out of 2 voting

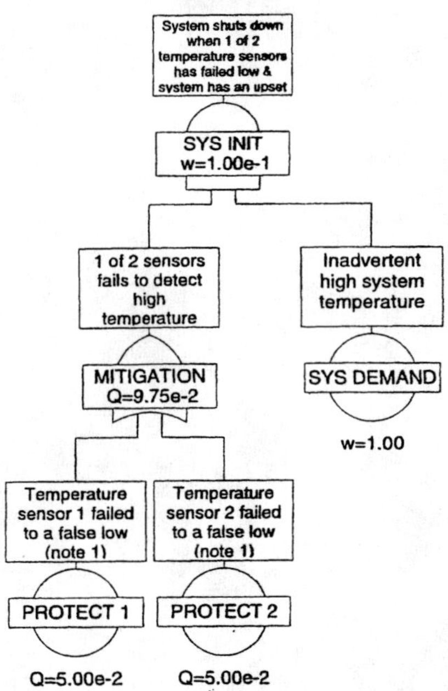

w - is a frequency value (upsets/yr)
Q - is a probability value

NOTES:
1. 5.0 e-2 = 0.59/yr x (2 mo/2 x yr/12 mo)
test interval = 2 mo

Figure A-4B: An unwanted accident with 2 out of 3 voting

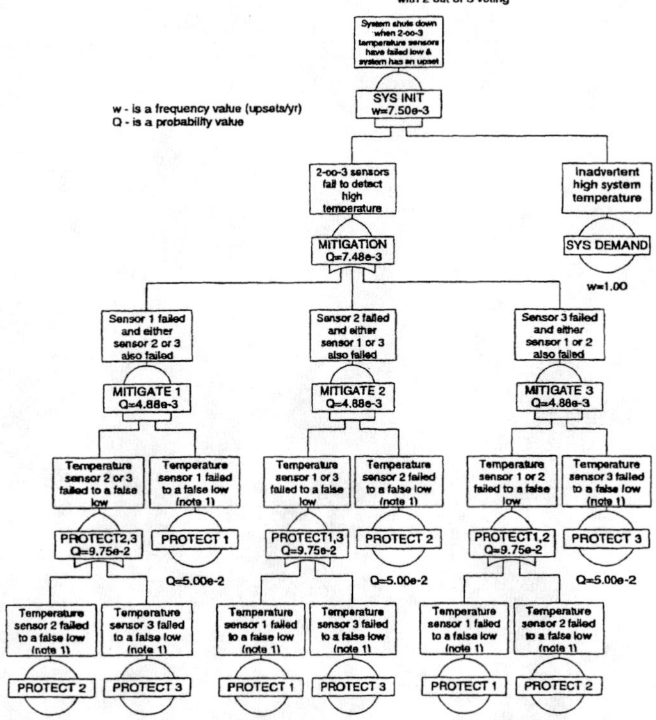

NOTES:
1. 5.0 e-2 = 0.59/yr x (2 mo/2 x yr/12 mo) test interval = 2 mo

Proceedings of
2001 ASME International Mechanical Engineering Congress and Exposition
November 11–16, 2001, New York, NY
SERA-Vol. 11

IMECE2001/SERA-24011

HUMAN ERROR ASSESSMENT AND DECISION MAKING FOR FISHING VESSELS USING ANALYTICAL HIERARCHY PROCESSING

A. Pillay & J. Wang

School of Engineering, Liverpool John Moores University,

Liverpool, Byrom Street, L3 3AF, UK.

Tel.: +44 151 231 2140

Fax: +44 151 298 2624

Email: engapill@livjm.ac.uk

ABSTRACT

An approach to integrate human error assessment and decision-making using the Analytical Hierarchy Processing (AHP) method is proposed. The aim of this approach is to reduce the probability of occurrence and severity of human error during the operational phase of a fishing vessel. It utilises AHP theory to rank the impacts of human error and further integrates the available control options (to minimise these errors) within the analysis. The result obtained from the analysis reflects the most favoured control option that will address all the possible human errors within the system to a satisfactory level. A test case, which considers the shooting operation of a beam trawler, is used to demonstrate the proposed approach. Each step involved in the shooting operation is assessed for its vulnerability to human error with respect to the equipment being operated and this captures the operator-machine interaction. The AHP method integrates the evaluation of solutions to reduce risk levels within the human error assessment and this reduces the complexity of the analysis that is present in traditional methods.

Introduction

The cost of shipping casualties is normally expressed in terms of insurance value. The report of the Institute of London Underwriters (ILU) for 1995 stated that 95 ships were lost during the year [ITSA, 1996; ILU, 1996]. In 1996, the ILU recorded 1,190 lives lost at sea and the ship classification society Det Norske Veritas (DNV) has estimated that accidents on board ships cost the industry around $US 10 billion a year [ILU, 1996; IMO, 1997]. It has been accepted that 80% of the accidents in the maritime industry is caused by human error. In the fishing vessel industry, Lloyd's Register of World Fleet Statistics 1998 notes that the average age of the world fleet of fish catching vessels over 100 GRT (Gross Registered Tonnage) is 20 years [International Transport Workers Federation (ITF), 1999]. This could be a contributing factor to the high level of human error on these vessels. These older vessels lack automation and modern safety devices, hence the safe operation of the vessels is highly dependent on the competency of the crew on board.

Human error has played a critical role in the causes of many major marine accidents. The officers and crew of the Herald of Free Enterprise set to sea with their bow doors open [Sheen N., 1987]. The crew and skipper of the Pescalanza and Sapphire did not close their watertight doors during heavy seas, which led to the sinking of the vessels by flooding [MAIB, 2000].

In these accidents, life, cargo and property had been lost due to the negligence and/or mistakes made by the operators of the system. Understanding errors and system failures are particularly important with respect to "high-consequence" systems. These are open systems whose behaviour has a significant effect not only on the system itself but also on the world outside the system. Hence, there is a need for an effective method to model the risks posed by human error in order to direct the limited resources to solutions that would reduce these risks.

Review of Human Error Assessment Methods

Engineers have developed a range of tools that can be used to represent and reason about the causes of major accidents [Leveson N., 1995]. For example, time-lines and fault trees have been recommended as analysis tools by a range of government and regulatory bodies. Unfortunately, these well-established techniques suffer from a number of limitations [Johnson C., 1998]. In particular, they cannot easily be used to represent and reason about the ways in which human errors and system failures interact during complex accidents [Hollnagel E., 1993].

Most methods for estimating human reliability were developed in nuclear power plants. Such methods include confusion matrix [Fullwood R.R. and Hall R.E., 1988; Gertman D.I. and Blackman H.S., 1994], expert estimation [Gertman D.I. and Blackman H.S., 1994], Time Reliability Curve (TRC) [Dougherty E.M. and Fragola J.R., 1988; Moieni P. et. al., 1994], MAaintenance Personnel Performance Simulation (MAPPS) [Fullwood R.R and Hall R.E., 1988; Gertman D.I. and Blackman H.S., 1994], Success Likelihood Index Method - Multi-Attribute Utility Decomposition (SLIM-MAUD) [Fullwood R.R. and Hall R.E., 1988; Gertman D.I. and Blackman H.S., 1994], sociotechnical assessment of human reliability [Gertman D.I. and Blackman H.S., 1994], Technique of Human Error Rate Prediction (THERP) [Dhillon B. S., 1986], Sandia Recovery Model (SRM), INTENT [Gertman D.I. et. al., 1992] and Operator Reliability Calculation and Assessment (ORCA). Of the methodologies given, most deal with misdiagnosis or non-response errors and time dependent probability estimates. The most commonly used techniques are THERP, utilising generic Human Error Probabilities (HEP) from various industries, and SLIM-MAUD, using importance weightings from experts.

Technique for Human Error Rate Prediction (THERP) provides a mechanism for modelling as well as quantifying human error. It starts off with a task analysis that describes the tasks to be performed by the crew, maintainers or operators. Together with the task descriptions, Performance-Shaping Factors (PSF) such as stress and time available are collected to modify probabilities. The task analysis is then graphically represented in Human Reliability Assessment (HRA) event trees. The HEP for the activities of the task or the branches are read and/or modified from the THERP tables as shown in [Gertman D.I. and Blackman H.S, 1994]. Details on the construction of HRA event trees and also, the COGnitive EveNt Tree (COGENT) to represent cognitive activities and errors associated with human performance are also given in the book. Gertman and Blackman also provide a summary of the steps to approach THERP, which was adapted from the NUREG/CR-1278 [Swain A. D. and Guttmann H. E., 1983]. THERP suffers from the following limitations [Reason J., 1990; White D., 1995]:

- It is difficult to represent the variability of human behaviour adequately.
- The technique assumes each task segment can be handled separately.
- It is difficult to combine human and equipment reliability values.
- It is difficult to identify inter-task dependencies.
- The technique is not appropriate for continuous tasks.
- The method does not determine motivation of the individual.
- Analysts have the tendency to model only errors that appear in databases.

AHP for human error assessment and decision making for fishing vessels

The methods available to quantify human error suffer from the difficulty associated with any attempt to construct quantitative, predictive models of human behaviour. The qualitative methods on the other hand, require multi-disciplinary teams to carry out an analysis and this is regarded as being resource intensive. The more recent HRA methods have included cognitive aspects of decision making and the "time" dimension. However, it has not yet captured the fundamental nature of the interaction between actions and machine responses [Cacciabue P.C. et. al., 1993]. These interactions lie in the mutual dynamic influence of the operator, the plant and the interfaces.

The use of AHP to evaluate human error on fishing vessels does not ignore small events or operations that are normally rationalised and eliminated as being not important in traditional methods. A chain of these small rationalisations results in a larger problem later. The AHP method looks at every event/operation and ranks them against each other to determine the importance of each event/operation over the other (without eliminating them from the analysis).

The use of the AHP method enables the solutions for each possible human error identified, to be integrated within the analysis. This is unlike the traditional methods, where the solutions to reduced the risk levels (posed by human errors) are evaluated in the first instance, and then a re-iteration of the whole analysis is performed (assuming the implementation of the solution) to confirm the risk reduction.

A Proposed Approach

The flowchart in figure 1 illustrates the steps involved in carrying out the proposed approach. This approach can be executed in seven distinct steps:

1. Describe operation - The system or operation under consideration is described in detail, highlighting all the equipment within the system that will be operated to achieve the desired objective of the defined operation.
2. Identify tasks to be carried out - Identify all tasks that are to be carried out to achieve the objective of the operation and present these tasks in the order that they should be carried out. This information can be represented by means of a flow chart. The order by which the tasks are carried out should reflect the normal safe working procedure of the system. To enable effective use of this information in the AHP phase, all tasks are annotated according to the equipment that are operated.
3. Determine operator behaviour - For each of the task identified in step 2, determine the required operators behaviour. Three types of behaviours are considered namely, skill-based, rule-based or knowledge-based behaviour.
4. Determine the probability of occurrence - Using a generic database, determine the probability that a human error might occur while carrying out the task specified in step 2. Use the information developed in step 3 to assign the probability of occurrence of the human error.
5. Determine the severity of occurrence - The severity of a human error should take into account the consequences of the error on the system, operation, environment and operator. This can be quantified in monetary terms or downtime.
6. Identify Risk Control Options (RCO) - Considering the system/operation under study; determine several options that could address the risks identified (associated with each task defined in step 2). The RCOs proposed should be carefully thought through and be feasible for implementation within the system/operation studied.
7. AHP analysis - Using the data gathered in steps 2, 4, 5 and 6, carry out the AHP analysis to determine the most favourable RCO. This RCO will address all the risks associated with tasks where human errors could manifest.

Step 7 (AHP analysis) involves 4 distinct steps, which are described here:

a) Set-up - Decision making criteria are generated, often by brainstorming or past experience. Hierarchical relationships are drawn between the criteria and are then represented in a matrix form.
b) Weighting - The matrices are filled with the criteria comparisons. The comparisons allow calculation of the criteria-weighting vector.
c) Ranking - The different RCOs are ranked on their ability to satisfy the various criteria.
d) Evaluation - The final solution ratings are then calculated using the ratings determined in step (c) and the weighting vector calculated in step (b).

The first task is to decide on the problem statement. This statement becomes Level One or the goal of the hierarchy and will be broken down into nested levels (Level Two). Level Two will comprise of the different elements needed to be considered to achieve the goal set in the problem statement. The elements in Level Two is further broken-down to represents the various constituents that make up or

belong to each specific element. The hierarchical structure is assumed to exist inherently in the problem considered and can be identified.
The hierarchy records the flow of detail from the problem statement (Goal) to broad issues (Level Two) and more specific levels (Level Three). While the concerns on a particular level are not equally important, they should be on the same order of magnitude. This feature in AHP allows decisions to be made involving different orders of magnitude criteria, by placing each criterion in its proper matrix in the objective hierarchy. Figure 2 shows an example of the hierarchy represented diagrammatically.

Once the hierarchy has been completed, matrices are constructed with the criteria labels on each axis. There will be one Level Two matrix and one matrix containing the sub-elements of each element. For example, the hierarchy sample in figure 2 will have one Level Two matrix and three Level Three matrices. These Level Three matrices may be broken down in finer detail where applicable. The two axes of the matrix will contain the names of the elements on the level being considered. For example, the Level Two matrix in figure 2 will have the form shown in figure 3. The levels below each of the Level Two elements would also be represented in a matrix form. Figure 4 shows an example for the Element 1 (constituent A) matrix. The complete representation of Element 1 would comprise of three matrices (as Element 1 has the constituents A, B and C).

Upon generating the matrices for all the elements, it must now be filled with the comparisons of the relative importance of the elements on the two axes. The comparisons are used to calculate the weighting vector that will give the relative importance of all the elements. The entire weighting vector is calculated from comparisons made between just two elements at a time. Table 1 shows the scale (1 to 9) proposed by [Saaty T.L, 1980] for indicating the relative importance between the elements.

Considering the example of the Level Two matrix in figure 3, and assuming that Element 1 is weekly more important than Element 2 and strongly more important than Element 3. Then, the matrix in figure 3 may be represented as seen in the matrix below:

$$\textit{Level Two} = \begin{bmatrix} 1 & 3 & 7 \\ 1/3 & 1 & 7/3 \\ 1/7 & 3/7 & 1 \end{bmatrix}$$

In the matrix, Element 1 is of equal importance with respect to itself, so 1 is placed in the upper left-hand corner. A consistent matrix formulation allows the remainder of the matrix to be completed given the information in the top row. Since the relationship is known between Element 1 and Element 2, and Element 1 and Element 3, the relationship between Element 2 and Element 3 can be determined. In this case the matrix entry for Element 2 versus Element 3 would contain 7/3. Similarly, the rest of the matrix can be computed using the formula $a_{ij} = a_{Ik} / a_{Ij}$.

The weighting vector is then determined to give the percentage of the total weight applied to each element. The first column in the Level Two matrix, (1, 1/3, 1/7) is normalised so that the sum of the entries is 1.0. The weighting vector of Element 1 will be given as 1/(1+1/3+1/7) = 0.667 or 67.8%. Similarly Elements 2 and 3 can be calculated to be 22.6% and 9.68%. The sum of all three weighting vectors should be equal to 100%.

The comparison process is repeated for all the matrices to be used in the analysis. The weighting vectors of the lower matrices will be normalised by the weight associated to the sub-elements' matrix, so that their total weight will equal that of the previous level (Level Two). For example, for Element 1 sub-element A_1, A_2, A_3, B_1, B_2, C_1, C_2 and C_3 will be given a total weight of 67.8%. All sub-elements are analysed in the same fashion to the lowest level possible and the results are normalised to reflect the weight of each sub-element in the hierarchy.

The next step is to generate the possible solution to achieve the problem statement/goal. Each solution is compared against each of the lowest level sub-elements. The possible solutions are assumed to reduce the likelihood of human error occurring and/or the possible consequences. The evaluation represents the "effectiveness" of the solution in controlling the risks. These evaluations (of the solutions) are recorded with a user defined numerical scale, as appropriate for the sub-elements. For any given element, a normalised score is determined for each solution by taking the assigned score (which may have units) and dividing it by the sum of the assigned scores across all of the solutions. This fraction is then multiplied by the weighting coefficient for the element. This will give a normalised score for each solution based on the element considered. These normalised results are then summed down the different elements in the matrix, to arrive at a final rating for each solution. The result of this series of operations is a weighted rating for each solution. The highest rated solution will best meet the problem statement (goal).

An Example

The purpose of this analysis is to address the high level of human errors that occur during the fishing operation on board fishing vessels. As an example, the initial shooting operation of the fishing nets is considered.

At the commencement of the voyage the beams are stowed port and starboard alongside and inboard of the bulwark rails. The cod ends are held by the Gilson wires up at the cod end lifting blocks with the netting hanging down to the chain mat that is beneath the beam. As soon as the vessel clears the harbour, the derricks are lowered to an angle of approximately 45 degrees. This reduces the top weight on the vessel, improving stability but importantly, it is to prevent the derricks from moving past vertical and falling inboard as the vessel rolls. On reaching the fishing grounds, the vessel stops and, working one side at a time, the derrick is raised sufficient to lift the beam and chain mat up and over the rail, the derrick being lowered back to 45 degrees on completion of the manoeuvre. While the cod ends are held by Gilson wire over the lifting block, the netting is paid overboard. Attached between the inboard end of the beam and the cod end lifting becket is a heavy rope, referred to as a 'lazy decky'. This rope is pulled to swing the beam around to bring it normal to the vessel side. The vessel moves ahead slowly and the Gilson wires are lowered slightly, sufficient to allow the cod ends to be swung over the rail; but still with the Gilson hooks attached in the lifting becket, the weight being carried by the cod ends lifting blocks. The crew, on each side, then takes the 'lazy decky' and makes it fast on a bulwark rail pin such that the weight of cod end is carried by the pin. The remainder of the 'lazy decky' lies in a bight on the deck up to the point where it goes over the rail to hang in a bight between the vessel and the beam. Once the weight of the cod end has been transferred to the rail pins, the Gilson hooks are released, the derricks are lowered fully outboard, and the vessel is brought up to speed. When the crew in the wheelhouse, either the skipper or the mate, is satisfied that the vessel is running straight and true, he signals to the crewman to release the 'lazy decky' ropes from the rail pins. The cod ends then stream astern with the netting stretched out. Warp is then paid out, typically 200 fathom for sea depth of 40 fathom. The complete operation can be represented diagrammatically as seen in figure 5.

Hierarchy set-up: The various tasks identified in figure are used to set up the hierarchy of elements. The goal of the analysis is determined to be the safe initial shooting operation. The elements in Level Two is set to be the probability of a human error occurring and the severity of human error. The sub-elements (Level Three) are determined by grouping the equipment that are operated i.e. derrick, vessel, lazy decky, net and Gilson. Each task carried out in relation to these equipment is considered within this level i.e. derrick 1, derrick 2, derrick 3, etc. The hierarchy can be represented diagrammatically as seen in figure 6.

Level Two matrix: The probability of occurrence and severity make up the two elements in Level Two as seen in figure 6. These two elements are compared against each other to determine the weighting vector of each element. The comparison scale in table 1 is used to determine the importance of the two elements. Considering the goal of the analysis, it is decided that both these elements are equally important to a safety assessment, hence, the Level Two matrix and the weighting vector is determined as:

$$Level\ \ Two = \begin{bmatrix} 1 & 1 \\ 1 & 1 \end{bmatrix},\ Weighting\ Vector = \begin{bmatrix} 0.5 \\ 0.5 \end{bmatrix}$$

Human error probability evaluation: First, the importance of each element (derrick, vessel, lazy decky, net and Gilson) is determined. Using the comparison scale in table 1, the matrix below is obtained for the probability importance of each element.

$$\textbf{Probability} = \begin{bmatrix} 1.00 & 7.00 & 3.00 & 9.00 & 5.00 \\ 0.14 & 1.00 & 0.43 & 1.29 & 0.71 \\ 0.33 & 2.33 & 1.00 & 3.00 & 1.67 \\ 0.11 & 0.78 & 0.33 & 1.00 & 0.56 \\ 0.20 & 1.40 & 0.60 & 1.81 & 1.00 \end{bmatrix}$$

The weighting vector and normalised vector is determined by considering the weighting vector obtained in the Level Two matrix and is shown here:

$$Weighting\ \ Vector = \begin{bmatrix} 0.5595 \\ 0.0799 \\ 0.1865 \\ 0.0622 \\ 0.1119 \end{bmatrix},\ Normalised\ Vector = \begin{bmatrix} 0.2798 \\ 0.0040 \\ 0.0933 \\ 0.0311 \\ 0.0560 \end{bmatrix}$$

In mathematical terms, the principle eigenvector is computed, and when normalised becomes the vector of priorities (weights). To reduce the excessive computing time needed to solve the problem exactly, and due to the results of complex numbers, a good estimate of that vector can be obtained by dividing the elements of each column in the comparison matrix by the sum of that column (i.e. normalise the column). The elements in each resulting row are added and the sum is divided by the number of the elements in the row.

The probability of human error is considered for each of the task carried out by determining the type of human behaviour required to carry out the task successfully. Using the generic human error data by Rasmussen [Rasmussen J., 1982], each task is assigned operator behaviour and the generic error probability. This data is then used to compare each task against the others to determine the Level Three matrix. The matrices for the probability of occurrence for each task are determined as follows:

$$Derrick = \begin{bmatrix} 1.00 & 1.00 & 1.00 & 0.11 \\ 1.00 & 1.00 & 1\ 00 & 0.11 \\ 1\ 00 & 1.00 & 1.00 & 0.11 \\ 9.00 & 9\ 00 & 9.00 & 1.00 \end{bmatrix} \quad Normalised\ \ Vector = \begin{bmatrix} 0.0233 \\ 0.0233 \\ 0.0233 \\ 0.2098 \end{bmatrix}$$

$$Vessel = \begin{bmatrix} 1.00 & 5.00 & 1.00 & 9.00 \\ 0.20 & 1.00 & 0.20 & 1.80 \\ 1.00 & 5.00 & 1.00 & 9.00 \\ 0.11 & 0.56 & 0.11 & 1.00 \end{bmatrix} \quad Normalised\ \ Vector = \begin{bmatrix} 0.0173 \\ 0.0035 \\ 0.0173 \\ 0.0019 \end{bmatrix}$$

$$Lazy\ \ Decky = \begin{bmatrix} 1.00 & 1.00 & 1.00 \\ 1.00 & 1.00 & 1.00 \\ 1.00 & 1.00 & 1.00 \end{bmatrix} \quad Normalised\ \ Vector = \begin{bmatrix} 0.0311 \\ 0.0311 \\ 0.0311 \end{bmatrix}$$

$$Net = \begin{bmatrix} 1.00 & 1.00 \\ 1.00 & 1.00 \end{bmatrix} \quad Normalised\ \ Vector = \begin{bmatrix} 0.0155 \\ 0.1155 \end{bmatrix}$$

$$Gilson = \begin{bmatrix} 1.00 & 0.20 \\ 5.00 & 1.00 \end{bmatrix} \quad Normalised\ \ Vector = \begin{bmatrix} 0.0093 \\ 0.0466 \end{bmatrix}$$

Human error severity evaluation: The importance of each element (derrick, vessel, lazy decky, net and Gilson) is determined using the comparison scale in table 1, the matrix below is obtained for the severity importance of each element. The weighting vector and normalised vector is determined by considering the weighting vector obtained in the Level Two matrix and is shown here:

$$Severity = \begin{bmatrix} 1.00 & 7.00 & 3.00 & 9.00 & 5.00 \\ 0.14 & 1.00 & 0.43 & 1.29 & 0.71 \\ 0.33 & 2.33 & 1.00 & 3.00 & 1.67 \\ 0.11 & 0.78 & 0.33 & 1.00 & 0.56 \\ 0.20 & 1.40 & 0.60 & 1.81 & 1.00 \end{bmatrix} \quad Weighting\ \ Vector = \begin{bmatrix} 0.5595 \\ 0.0799 \\ 0.1865 \\ 0.0622 \\ 0.1119 \end{bmatrix}$$

$$Normalised\ \ Vector = \begin{bmatrix} 0.2798 \\ 0.0040 \\ 0.0933 \\ 0.0311 \\ 0.0560 \end{bmatrix}$$

The matrices for the severity of the consequences of human error for each task is determined as follows:

$$Derrick = \begin{bmatrix} 1.00 & 7.00 & 3.00 & 5.00 \\ 0.14 & 1.00 & 0.43 & 0.71 \\ 0.33 & 2.33 & 1.00 & 1.67 \\ 0.20 & 1.40 & 0.60 & 1.00 \end{bmatrix} \quad Normalised\ \ Vector = \begin{bmatrix} 0.1669 \\ 0.0238 \\ 0.0556 \\ 0.0334 \end{bmatrix}$$

$$Vessel = \begin{bmatrix} 1.00 & 0.20 & 0.33 & 0.11 \\ 5.00 & 1.00 & 1.67 & 0.56 \\ 3.00 & 0.60 & 1.00 & 0.33 \\ 9.00 & 1.80 & 3.00 & 1.00 \end{bmatrix} \quad Normalised\ Vector = \begin{bmatrix} 0.0022 \\ 0.0111 \\ 0.0067 \\ 0.0200 \end{bmatrix},$$

$$Lazy\ Decky = \begin{bmatrix} 1.00 & 0.11 & 0.33 \\ 9.00 & 1.00 & 3.00 \\ 3.00 & 0.33 & 1.00 \end{bmatrix} \quad Normalised\ Vector = \begin{bmatrix} 0.0072 \\ 0.0646 \\ 0.0215 \end{bmatrix}$$

$$Net = \begin{bmatrix} 1.00 & 5.00 \\ 0.20 & 1.00 \end{bmatrix} \quad Normalised\ \ Vector = \begin{bmatrix} 0.0259 \\ 0.0052 \end{bmatrix},$$

$$Gilson = \begin{bmatrix} 1.00 & 0.11 \\ 9.00 & 1.00 \end{bmatrix} \quad Normalised\ \ Vector = \begin{bmatrix} 0.0056 \\ 0.0504 \end{bmatrix}$$

Risk Control Options (RCO): Several viable Risk Control Options (RCO) are generated in order to reduce the level of risks posed by human errors during the initial shooting operation. These risk control options are evaluated for their effectiveness against each of the operator tasks identified. For this example, an arbitrary scale (1 to 10) is used to compare each RCO, 1 being not effective and 10 being most effective. When assigning a score on the 1 to 10 scale, several factors are considered, such as, cost, ease of implementation, efficiency, time before solution becomes effective, etc. Six RCOs' have been identified to reduce the probability and severity of human errors of the initial shooting operation. These RCOs include:

RCO 1 - Training of crew

RCO 2 - Redesign system
RCO 3 - Incorporate additional interlocks
RCO 4 - Change operating procedures
RCO 5 - Additional crewing
RCO 6 - Install warning devices

The matrices for the effectiveness of each RCO in reducing the probability of occurrence are presented in the form as seen in figure 7. Similarly all tasks are compared with the different RCOs and their values normalised as seen below.

RCO evaluation to reduce probability of occurrence

$$Derrick = \begin{bmatrix} 6 & 1 & 7 & 3 & 4 & 9 \\ 6 & 2 & 8 & 5 & 1 & 7 \\ 6 & 1 & 7 & 3 & 4 & 9 \\ 6 & 6 & 9 & 7 & 1 & 9 \end{bmatrix}$$

$$Normalised\ results = \begin{bmatrix} 0.0047 & 0.0008 & 0.0054 & 0.0023 & 0.0031 & 0.0070 \\ 0.0048 & 0.0016 & 0.0064 & 0.0040 & 0.0008 & 0.0056 \\ 0.0047 & 0.0008 & 0.0054 & 0.0023 & 0.0031 & 0.0070 \\ 0.0331 & 0.0331 & 0.0497 & 0.0386 & 0.0055 & 0.0497 \end{bmatrix}$$

$$Vessel = \begin{bmatrix} 8 & 1 & 3 & 2 & 4 & 7 \\ 7 & 1 & 2 & 2 & 4 & 7 \\ 7 & 1 & 3 & 2 & 4 & 7 \\ 9 & 5 & 3 & 2 & 6 & 8 \end{bmatrix}$$

$$Normalised\ results = \begin{bmatrix} 0.0055 & 0.0007 & 0.0021 & 0.0014 & 0.0028 & 0.0048 \\ 0.0011 & 0.0002 & 0.0003 & 0.0003 & 0.0006 & 0.0011 \\ 0.0050 & 0.0007 & 0.0022 & 0.0014 & 0.0029 & 0.0050 \\ 0.0005 & 0.0003 & 0.0002 & 0.0001 & 0.0003 & 0.0005 \end{bmatrix}$$

$$Lazy\ Decky = \begin{bmatrix} 4 & 3 & 7 & 2 & 6 & 9 \\ 5 & 2 & 8 & 3 & 6 & 8 \\ 6 & 4 & 8 & 3 & 7 & 8 \end{bmatrix}$$

$$Normalised\ results = \begin{bmatrix} 0.0040 & 0.0030 & 0.0070 & 0.0020 & 0.0060 & 0.0090 \\ 0.0049 & 0.0019 & 0.0078 & 0.0029 & 0.0058 & 0.0078 \\ 0.0052 & 0.0035 & 0.0069 & 0.0026 & 0.0060 & 0.0069 \end{bmatrix}$$

$$Net = \begin{bmatrix} 2 & 2 & 8 & 3 & 4 & 10 \\ 3 & 3 & 7 & 3 & 4 & 10 \end{bmatrix}$$

$$Normalised\ results = \begin{bmatrix} 0.0011 & 0.0011 & 0.0043 & 0.0016 & 0.0021 & 0.0054 \\ 0.0016 & 0.0016 & 0.0036 & 0.0016 & 0.0021 & 0.0052 \end{bmatrix}$$

$$Gilson = \begin{bmatrix} 5 & 3 & 6 & 5 & 6 & 7 \\ 7 & 3 & 10 & 6 & 6 & 10 \end{bmatrix}$$

$$Normalised\ results = \begin{bmatrix} 0.0015 & 0.0009 & 0.0017 & 0.0015 & 0.0017 & 0.0020 \\ 0.0078 & 0.0033 & 0.0111 & 0.0067 & 0.0067 & 0.0111 \end{bmatrix}$$

RCO evaluation to reduce severity of occurrence

$$Derrick = \begin{bmatrix} 6 & 1 & 5 & 5 & 4 & 7 \\ 6 & 2 & 7 & 5 & 4 & 7 \\ 6 & 1 & 5 & 6 & 4 & 7 \\ 6 & 4 & 6 & 7 & 5 & 9 \end{bmatrix}$$

$$Normalised\ results = \begin{bmatrix} 0.0358 & 0.0060 & 0.0298 & 0.0298 & 0.0238 & 0.0417 \\ 0.0046 & 0.0015 & 0.0054 & 0.0038 & 0.0031 & 0.0054 \\ 0.0115 & 0.0019 & 0.0096 & 0.0115 & 0.0077 & 0.0134 \\ 0.0054 & 0.0036 & 0.0054 & 0.0063 & 0.0045 & 0.0081 \end{bmatrix}$$

$$Vessel = \begin{bmatrix} 5 & 1 & 7 & 5 & 4 & 7 \\ 5 & 1 & 7 & 5 & 4 & 7 \\ 5 & 1 & 7 & 5 & 4 & 7 \\ 9 & 5 & 8 & 7 & 6 & 8 \end{bmatrix}$$

$$Normalised\ results = \begin{bmatrix} 0.0004 & 0.0001 & 0.0005 & 0.0004 & 0.0003 & 0.0005 \\ 0.0019 & 0.0004 & 0.0027 & 0.0019 & 0.0015 & 0.0027 \\ 0.0011 & 0.0002 & 0.0016 & 0.0011 & 0.0009 & 0.0016 \\ 0.0042 & 0.0023 & 0.0037 & 0.0033 & 0.0028 & 0.0037 \end{bmatrix}$$

$$Lazy\ Decky = \begin{bmatrix} 4 & 3 & 7 & 6 & 6 & 7 \\ 5 & 2 & 8 & 6 & 6 & 8 \\ 6 & 4 & 8 & 7 & 7 & 8 \end{bmatrix}$$

$$Normalised\ results = \begin{bmatrix} 0.0009 & 0.0007 & 0.0015 & 0.0013 & 0.0013 & 0.0015 \\ 0.0092 & 0.0037 & 0.0148 & 0.0111 & 0.0111 & 0.0148 \\ 0.0032 & 0.0022 & 0.0043 & 0.0038 & 0.0038 & 0.0043 \end{bmatrix}$$

$$Net = \begin{bmatrix} 2 & 2 & 8 & 6 & 5 & 8 \\ 3 & 3 & 7 & 6 & 5 & 8 \end{bmatrix}$$

$$Normalised\ results = \begin{bmatrix} 0.0017 & 0.0017 & 0.0067 & 0.0050 & 0.0042 & 0.0067 \\ 0.0005 & 0.0005 & 0.0011 & 0.0010 & 0.0008 & 0.0013 \end{bmatrix}$$

$$Gilson = \begin{bmatrix} 5 & 3 & 6 & 5 & 6 & 7 \\ 7 & 3 & 7 & 6 & 6 & 8 \end{bmatrix}$$

$$Normalised\ results = \begin{bmatrix} 0.0009 & 0.0005 & 0.0010 & 0.0009 & 0.0010 & 0.0012 \\ 0.0095 & 0.0041 & 0.0095 & 0.0082 & 0.0082 & 0.0109 \end{bmatrix}$$

Summary of results

The results obtained above are collated to determine the best RCO. Tables 2 and 3 show the summary of these results obtained in percentage. Each of these tables represents 50% of the weight (as the RCO evaluation has been normalised) of the elements in Level Two of the hierarchy. The final ranking of the RCOs is achieved by adding the final ratings of each these tables for the respective RCOs. Table 4 shows the final results obtained for this analysis. From this table, it can be determined that the best control option to reduce the probability of occurrence and the severity (of human error) during the initial shooting operation is RCO 6. The results entail that by installing various warning and indication devices onto/for the equipment used for the initial shooting operation on a fishing vessel, the level of human error can be reduced and safe operation can be achieved.

Conclusion

Human errors on fishing vessels have contributed to a great number of accidents in the past and almost 20% of all accidents on these vessels are caused by negligence/carelessness of the crew. As such, the ultimate aim for carrying out a human error assessment on fishing vessel is to determine the best method by which accidents caused by these errors can be reduced. This would entail decreasing the risk level by either reducing the probability of a human error occurring or the severity of the consequences.

This paper proposes a method using AHP to achieve this aim. The approach integrates the risk control option within the human error assessment framework to determine the best option for the identified hazards. The advantages of using the proposed approach for fishing vessels include:

- The use of a flexible modeling and measurement approach to evaluation.
- The application of structure to facilitate decision making through the use of a model which imposes strict independence, ordinality, or homogeneity of preferences.
- Allowing the decision-maker to arrive at consistent and objective evaluations.
- The simplicity of the use of the model.

- The confidence that all human errors identified are evaluated (without being omitted by rationalisation) in the decision making process.
- The interaction between operator and machine is captured within the analysis.
- Precision is not forced (subjective method of evaluation).
- Does not require a large group (multi-disciplinary) of analyst.

In this paper, only human errors are considered in the analysis. However, this can be extended to include failures induced by other causes, such as machinery failure. As such, a comprehensive safety assessment can be completed using the AHP method.

References

1. Cacciabue P.C., Carpignano A. and Vivalda C. (1993), "A dynamic reliability technique for error assessment in man-machine systems", International Journal of Man-Machine Studies, Vol.38, pp. 403-428.
2. Dhillon B. S. (1986), Human reliability with human factors, Pergamon Press Inc., New York.
3. Dougherty E.M. and Fragola J.R. (1988), Human reliability analysis: A systems engineering approach with nuclear power plant applications, John Wiley & Sons Inc., Chichester, U.K.
4. Fullwood R.R. and Hall R.E. (1988), Probabilistic risk assessment in the nuclear power industry, 1st Ed. Pergamon Press.
5. Gertman D.I. and Blackman H.S (1994), Human reliability & safety analysis data handbook, 1st Ed. John Wiley & Sons Inc.
6. Gertman D.I., Blackman H.S., Haney L.N., Seidler K.S. and Hahn H.A. (1992), "INTENT : A method for estimating human error probabilities for decision based errors", Reliability Engineering and System Safety, Vol. 35, pp. 127-136.
7. Hollnagel E. (1993), "The phenotype of erroneous actions", International Journal of Man-Machine Studies, 39, pp. 1-32.
8. ILU (1996), Extract from database 1989-1996, The Institute of London Underwriters, London, United Kingdom.
9. IMO (1997), "Optimum maritime safety demands a focus on people", World Maritime Day 1997.
10. International Transport Workers Federation (ITF) (1999), "Troubled Waters: 1999 CSD thematic review (Oceans and seas)", March 1999, ISBN 0-901969-22-2.
11. International Transport Workers Federation (ITF) (1999), Troubled waters: 1999 CSD thematic review (Oceans and seas), March, ISBN 0-901969-22-2.
12. ITSA (International Transport Safety Association) (1996), "Serious ship accidents rose in first quarter of 1996", ITSA Report, Vol.2, No.4.
13. Johnson C. (1998), "Representing the impact of time on human error and system failure", Interacting with Computers, Vol.11, pp. 53-86.
14. Leveson N. (1995), System safety and computers, Addison Wesley, Readiing, Massachusetts, USA.
15. MAIB (2000), "Safety digest: Lessons from marine accident reports", 1/2000, Marine Accident Investigation Branch, DETR, UK.
16. Moieni P., Spurgin A. J. and Singh A. (1994), "Advances in human reliability analysis methodology. Part I: Frameworks, models and data", Reliability Engineering and System Safety, Vol. 44, pp. 27-55.
17. Rasmussen J. (1982), "Human errors: A taxonomy for describing human malfunction in industrial installations", Journal of Occupational Accidents, Vol. 4, pp. 311-333.
18. Reason J. (1990), Human error, Cambridge University Press, Cambridge.
19. Saaty T.L. (1980), The Analytic Hierarchy Process, McGraw Hill Company, New York.
20. Sheen N. (1987), "Formal investigations into the sinking of the MV Herald of Free Enterprise", UK Department of Transport, Report of Court 8074, Her Majesty's Stationery Office.
21. Swain A. D. and Guttmann H. E. (1983), "Handbook of human reliability analysis with emphasis on nuclear power plant applications", NUREG/CR-1278, August.
22. White D. (1995), "Application of system thinking to risk management: A review of the literature", Journal of Management Decision, Vol.33, No.10, pp. 35-45.

1	Both elements of equal importance			
3	Left weakly more important than top		1/3	Top weakly more important than left
5	Left moderately more important than top		1/5	Top moderately more important than left
7	Left strongly more important than top		1/7	Top strongly more important than left
9	Left absolutely more important than top		1/9	Top absolutely more important than left

Table 1 Comparison scale

	Derrick	Vessel	Lazy Decky	Net	Gilson	Total rating
RCO 1	*4.73%*	*1.22%*	*1.40%*	*0.26%*	*0.92%*	***8.53%***
RCO 2	*3.63%*	*0.19%*	*0.84%*	*0.26%*	*0.42%*	***5.34%***
RCO 3	*6.70%*	*0.47%*	*2.17%*	*0.79%*	*1.28%*	***11.42%***
RCO 4	*4.73%*	*0.32%*	*0.75%*	*0.32%*	*0.81%*	***6.94%***
RCO 5	*1.25%*	*0.66%*	*1.79%*	*0.42%*	*0.84%*	***4.97%***
RCO 6	*6.93%*	*1.14%*	*2.37%*	*1.05%*	*1.31%*	***12.81%***

Table 2 Summary of results for probability element

	Derrick	Vessel	Lazy Decky	Net	Gilson	Total rating
RCO 1	*5.73%*	*0.76%*	*1.33%*	*0.22%*	*1.04%*	***9.08%***
RCO 2	*1.30%*	*0.30%*	*0.65%*	*0.22%*	*0.46%*	***2.93%***
RCO 3	*5.02%*	*0.85%*	*2.06%*	*0.78%*	*1.06%*	***9.77%***
RCO 4	*5.15%*	*0.67%*	*1.61%*	*0.60%*	*0.90%*	***8.93%***
RCO 5	*3.91%*	*0.55%*	*1.61%*	*0.50%*	*0.92%*	***7.50%***
RCO 6	*6.87%*	*0.85%*	*2.06%*	*0.80%*	*1.21%*	***11.79%***

Table 3 Summary of results for severity element

	Derrick	Vessel	Lazy Decky	Net	Gilson	Total rating
RCO 1	*10.46%*	*1.98%*	*2.74%*	*0.48%*	*1.96%*	***17.61%***
RCO 2	*4.93%*	*0.49%*	*1.49%*	*0.48%*	*0.88%*	***8.27%***
RCO 3	*11.72%*	*1.33%*	*4.23%*	*1.57%*	*2.34%*	***21.19%***
RCO 4	*9.88%*	*0.99%*	*2.36%*	*0.91%*	*1.72%*	***15.87%***
RCO 5	*5.16%*	*1.21%*	*3.40%*	*0.92%*	*1.76%*	***12.46%***
RCO 6	*13.80%*	*1.99%*	*4.43%*	*1.85%*	*2.53%*	***24.60%***

Table 4 Final ranking of RCO

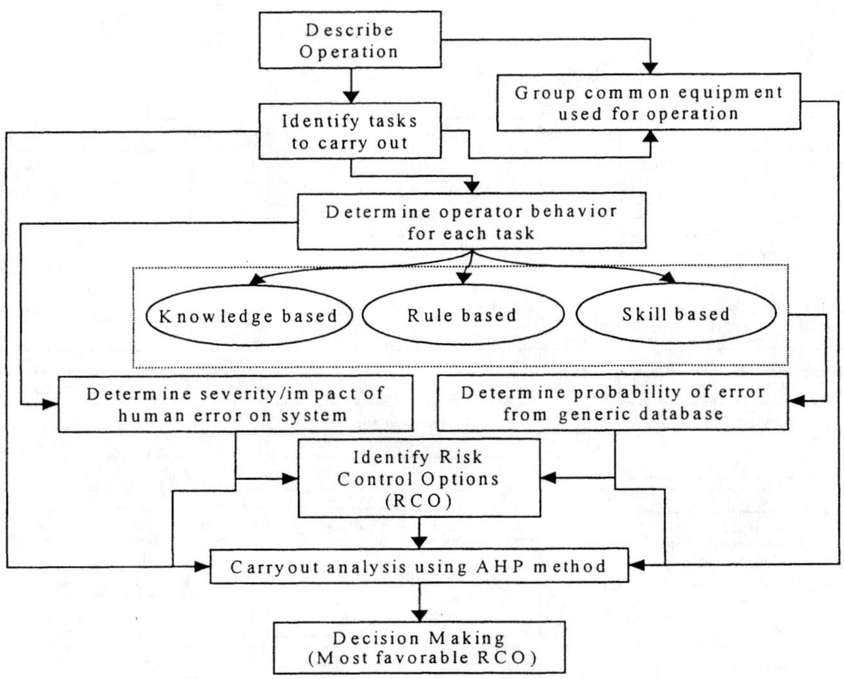

Figure 1 Flowchart of the proposed approach

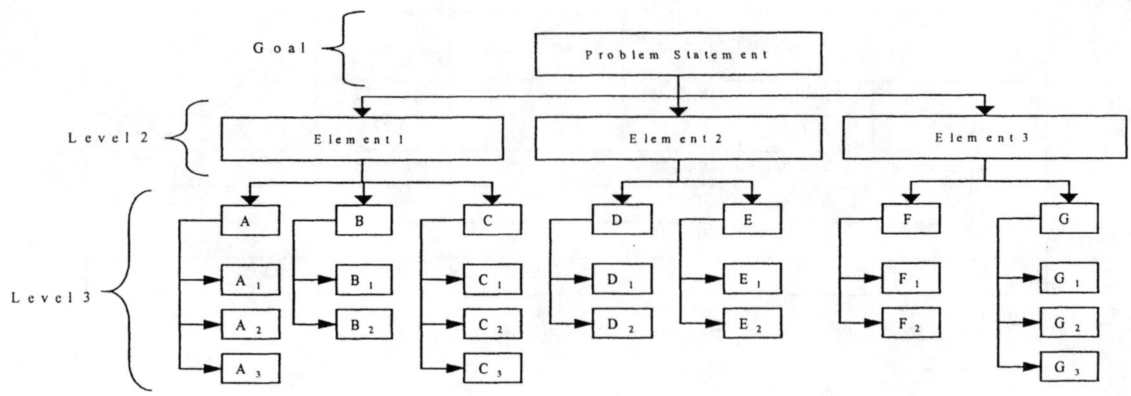

Figure 2 Example of hierarchy levels

Level Two	Element 1	Element 2	Element 3
Element 1	EL_{11}	EL_{12}	EL_{13}
Element 2	EL_{21}	EL_{22}	EL_{23}
Element 3	EL_{31}	EL_{32}	EL_{33}

Figure 3 Example of Level Two matrix

Element 1	A_1	A_2	A_3
A_1	A_{11}	A_{12}	A_{13}
A_2	A_{21}	A_{22}	A_{23}
A_3	A_{31}	A_{32}	A_{33}

Figure 4 Example of Level Three matrix

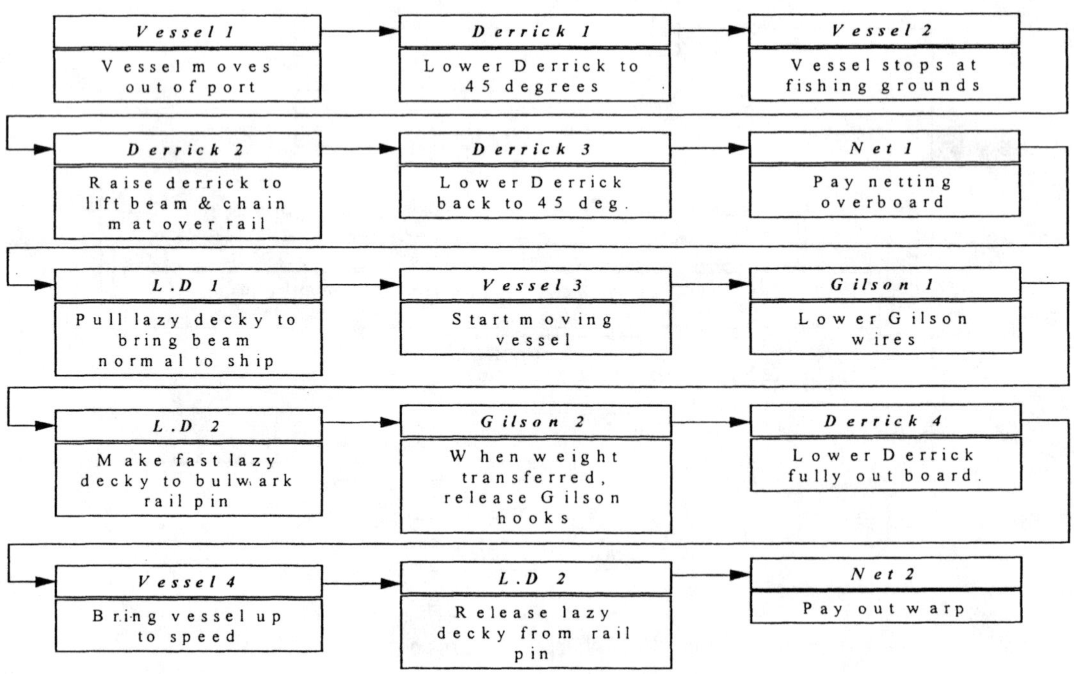

Figure 5 Diagrammatic representation of initial shooting operation

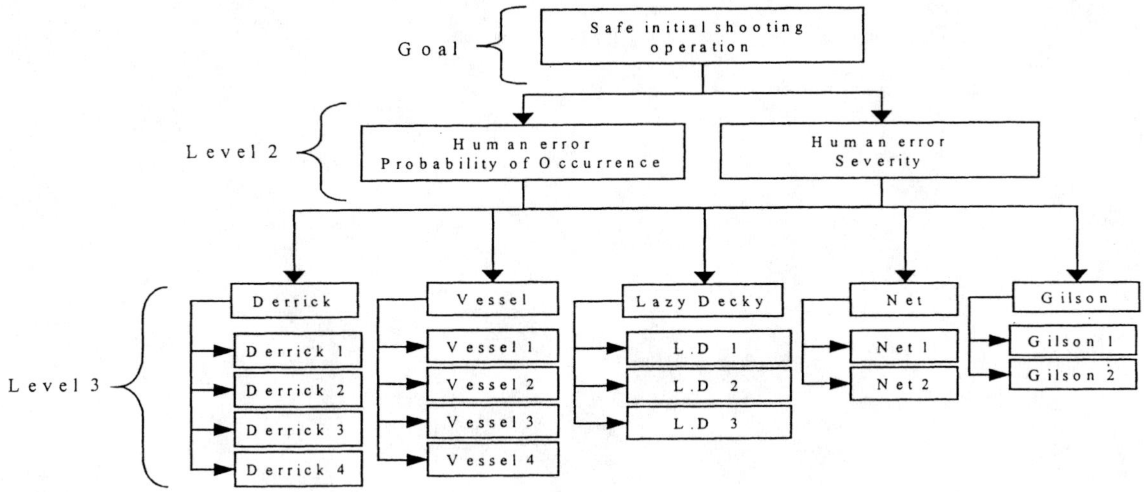

Figure 6 Initial shooting operation hierarchy levels

Derrick	RCO 1	RCO 2	RCO 3	RCO 4	RCO 5	RCO 6
Derrick 1						
Derrick 2						
Derrick 3						
Derrick 4						

Figure 7 RCO matrix

Proceedings of
2001 ASME International Mechanical Engineering Congress and Exposition
November 11–16, 2001, New York, NY
SERA-Vol. 11

IMECE2001/SERA-24012

FRICTION-RELATED ASPECTS OF HUMAN PUSHING

Cheryl A. Pattin, Ph.D.
Senior Mechanical and Biomechanical Engineer, Triodyne, Inc.
Ralph L. Barnett
Professor, Mechanical and Aerospace Engineering
Illinois Institute of Technology and Chairman, Triodyne, Inc.

5950 W. Touhy Ave.
Niles, IL 60714-4610 USA
(847) 677-4730 (voice)
cherylp@triodyne.com

ABSTRACT

Human pushing capability affects performance in the areas of crowd and animal control, the security of locked doors and railings, the removal of tree stumps and entrenched vehicles, the maneuvering of furniture, and in athletic pursuits such as football or wrestling. Pushing capability is a function of both individual characteristics (e.g., muscular strength, body weight, and the distribution of that weight) and attributes of the pushing task itself (e.g., footwear/floor and torso/object friction). Mathematical relationships among these factors are developed. Using these relationships, the maximum push that can be generated without slip were calculated for the following two conditions: (1) LEAN only, with the torso about to slip down the pushed object and feet about to slip along the floor and (2) PUSH along the body contact line with the torso about to slip up the pushed object and the feet about to slip along the floor.

INTRODUCTION

Horizontal pushing forces are generated by leaning against or actively pushing against an object with feet planted on the ground (or floor) and with some portion of the upper body in contact with a vertically projecting surface (Figs. 1a, 1b). Sometimes the effective body length may be altered, as when the knees are bent or the entire back presses against a wall (Fig. 1c). In this development the body will be considered as a line along the effective body length with ends maintaining single point contacts with the horizontal and vertical contact surfaces. Single point contacts resist only compressive and tangential loads, not moments. As a final approximation, any lengthening of the body between contact points will be neglected and the axial force generated by the body will be assumed to act along the line defined by the two contact points.

Previous estimates of pushing strength have been obtained for use in military standards (MIL-STD-1472D, 1989). Prior studies have also addressed human pushing and pulling capability with regard to metabolic cost and muscle moments (individual characteristics which are not the subject of this paper). Some of these results are presented and discussed in Chaffin, Andersson, and Martin (1999).

The nomenclature used in this paper and a definition of terms is provided at the end of this paper.

FREE BODY EQUATIONS

We idealize the human body during leaning and pushing as a rigid body with point contacts, incapable of sustaining moments, at the wall and floor. For the remainder of this paper, we will use the term floor to mean the horizontal surface and wall to mean the vertical surface of the object to be moved or acted upon. We will also use the term torso to represent that portion of the body in contact with the wall and feet for that portion of the body in contact with the floor. The body may be viewed as a series of distributed individual weights W_i at length a_i along the body from foot contact with the floor or ground to shoulder or torso contact with the vertical object to be pushed (a wall, for example) as shown in Figure 2.

These individual weights can be summed to obtain an equivalent body weight, W, acting at a point $\bar{a}/L$ along the body from the feet. Figure 3 shows the resulting free body diagram for the body with W and $\bar{a}/L$ calculated using equations 1 and 2.

$$W = \sum_{i=1}^{n} W_i \tag{1}$$

$$\bar{a} = \sum_{i=1}^{n} \left(\frac{W_i}{W} \right) a_i \tag{2}$$

The rigid leaning or pushing body may be modeled as shown in Figure 3 and the force and moment equations developed.

$$\sum F_x = 0 = H_t + H_b \tag{3}$$

$$\sum F_y = 0 = -W + V_t + V_b \tag{4}$$

$$\sum M_Z = 0 = (W \times \bar{a} \times \cos\alpha) - (V_t \times L \times \cos\alpha) - (H_t \times L \times \sin\alpha) \tag{5}$$

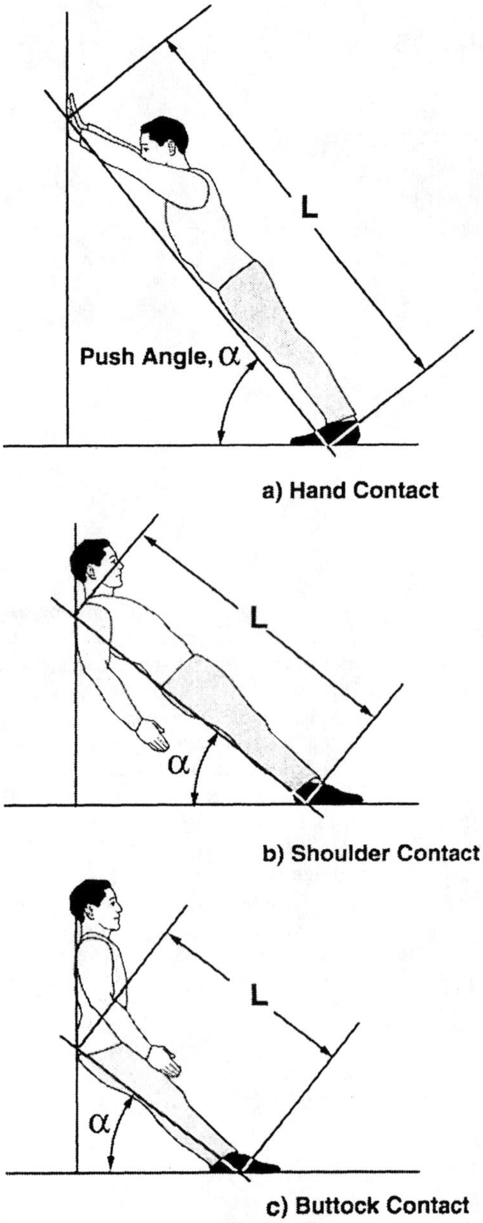

Figure 1. Pushing configurations

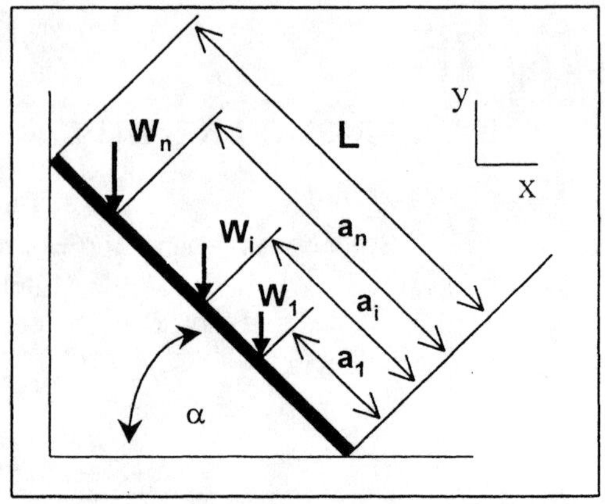

Figure 2. Distributed individual weights for body leaning or pushing against a wall.

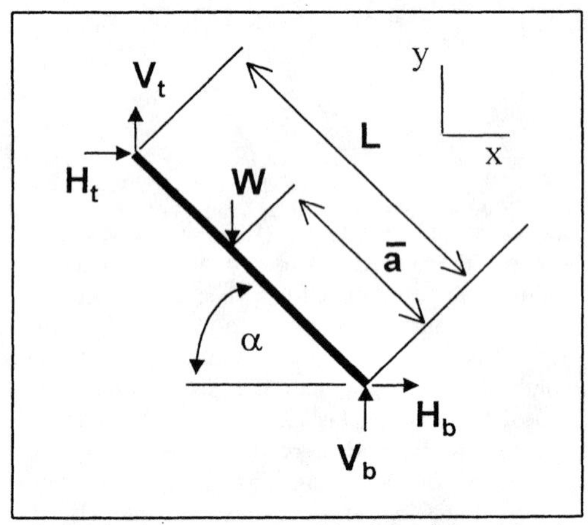

Figure 3. Equivalent concentrated body weight and point of action and free body diagram for leaning or pushing body diagrammed in Figure 2.

Equations 3 to 5 may be used to obtain expressions for 3 of the 4 unknown variables in this problem. The horizontal force H_t is left as an unknown variable.

$$V_t = W\left(\frac{\bar{a}}{L}\right) - H_t \times \tan\alpha \tag{6}$$

$$V_b = W\left(1 - \frac{\bar{a}}{L}\right) + H_t \times \tan\alpha \tag{7}$$

$$H_b = -H_t \tag{8}$$

$$H_t = \text{unknown} = \text{Horizontal Force} \tag{9}$$

HORIZONTAL FORCE GENERATED BY PASSIVE LEANING

We consider the case where the torso is about to slip down along the wall and the feet are about to slip along the floor to the right. In this case:

- the vertical reaction force at the wall, V_t, acts upward (+y) to resist incipient torso slip down the wall and
- the horizontal force at the floor, H_b, acts to the left (-x) to prevent incipient footwear slip to the right.

This loading condition will be referred to as **LEAN**.

For LEAN, the reaction forces are related by the friction coefficients between feet and floor (μ_b) and shoulder and wall (μ_t) in the following manner :

$$V_t = \mu_t H_t \tag{10}$$

$$H_b = -\mu_b V_b \tag{11}$$

Substituting equations 10 and 11 into equations 6-8 above, we obtain the following results:

The critical angle, α^{LEAN}, at which incipient slip occurs at both the torso-wall and footwear-floor interfaces is:

$$\alpha^{LEAN} = \tan^{-1}\left(\frac{\bar{a}}{L}\left(\mu_t + \frac{1}{\mu_b}\right) - \mu_t\right) \tag{12}$$

The horizontal force, F^{LEAN}, generated by a leaning body with incipient slip at the feet and torso is:

$$F^{LEAN} = W\frac{\mu_b}{1+\mu_b\mu_t} \tag{13}$$

Note that the magnitude of the horizontal force generated by passive leaning (F^{LEAN}) is a function of only the body weight, W, and the friction coefficients (μ_b and μ_t). The angle at which incipient slip will occur at both the wall and floor during lean (α^{LEAN}) is a function of the weight distribution ($\bar{a}/L$) and the friction coefficients (μ_b and μ_t). It should be pointed out that equation 12 above for pure LEAN is identical to that governing the static behavior of straight or extension ladders (Barnett, 1996). As an example, if $\mu_t = 0.25$, $\mu_b = 0.75$, and $\bar{a}/L = 0.8$, $F^{LEAN}/W = 0.63$ and $\alpha^{LEAN} = 45.5$ degrees.

HORIZONTAL FORCE GENERATED BY AXIAL THRUST

We consider the case where axial thrust is generated along the long axis of the body sufficient to cause the torso to almost slip vertically up along the wall and the feet to slip to the right. This case represents the maximum horizontal force that can be generated using axial thrust along the body. In this case:

- the vertical reaction force at the wall, V_t, acts downward (-y) to resist incipient torso slip up the wall and
- the horizontal force at the floor, H_b, acts to the left (-x) to prevent slip to the right.

This loading condition will be referred to as **PUSH**.

For PUSH, the reaction forces are related by the friction coefficients between feet and floor (μ_b) and shoulder and wall (μ_t) in the following manner :

$$V_t = -\mu_t H_t \tag{14}$$

$$H_b = -\mu_b V_b \tag{15}$$

Substituting equations 14 and 15 into equations 6-8 above, we obtain the following results:

The critical angle, α^{PUSH}, at which incipient slip occurs at both the torso-wall and footwear-floor interfaces is:

$$\alpha^{PUSH} = \tan^{-1}\left(\frac{\bar{a}}{L}\left(\frac{1}{\mu_b} - \mu_t\right) + \mu_t\right) \tag{16}$$

The horizontal force, F^{PUSH}, generated by a pushing body with incipient slip at the feet and torso is:

$$F^{PUSH} = W\frac{\mu_b}{1-\mu_b\mu_t} \tag{17}$$

As an example, if $\mu_t = 0.25$, $\mu_b = 0.75$, and $\bar{a}/L = 0.8$, $F^{PUSH}/W = 0.92$ and $\alpha^{PUSH} = 48.2$ degrees.

PUSH AND LEAN AS FUNCTIONS OF FEET/FLOOR AND TORSO/WALL FRICTION

The above relations (equations 12, 13, 16, and 17) may be examined in terms of the effects of μ_b and μ_t.

Figure 4 is a plot of F^{LEAN} / W and F^{PUSH} / W plotted as a continuous function of foot/floor friction (μ_b) for selected values of torso/wall friction (μ_t). Note that the maximum horizontal push during PUSH exceeds that for LEAN. Note also that the LEAN force never exceeds body weight (W). Figure 5 shows critical angles for $\bar{a}/L = 0.8$ for lean (α^{LEAN}) and push (α^{PUSH}) for the same values of μ_b and μ_t . Note that some purchase on the floor is required to generate horizontal push or lean or to sustain a non-vertical orientation.

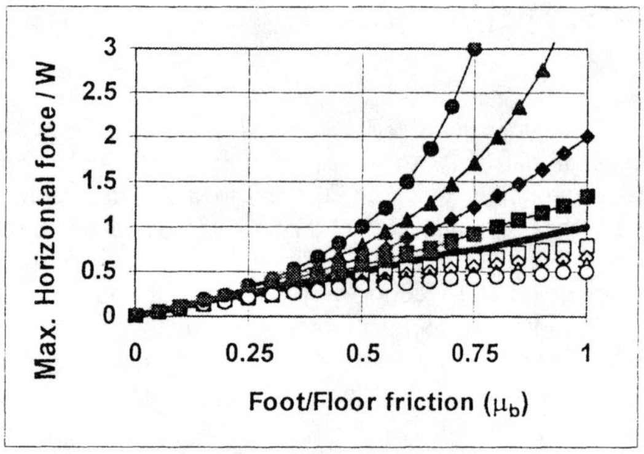

Figure 4. Horizontal forces generated by LEAN and PUSH normalized by body weight (W) as a function of μ_b for selected values of μ_t.

Critical Angles for $\bar{a}/L = 0.8$

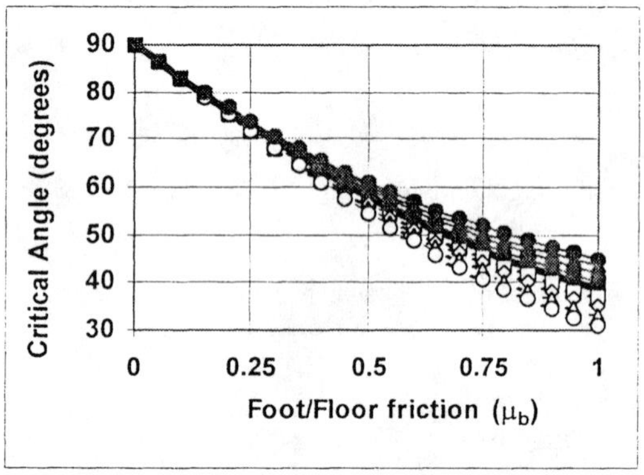

Figure 5. Critical slip angles in LEAN and PUSH as a function of μ_b for selected values of μ_t.

Legend for Figures 4 and 5:

F^{PUSH} Filled symbols F^{LEAN} Open Symbols

$\mu_t = 0.0$ Solid line
$\mu_t = 0.25$ Square icons
$\mu_t = 0.5$ Diamond icons
$\mu_t = 0.75$ Triangle icons
$\mu_t = 1.0$ Circle icons

Figure 6 is a plot of F^{LEAN} / W and F^{PUSH} / W plotted as a continuous function of torso/wall friction (μ_t) for selected values of foot/floor friction (μ_b). Figure 7 shows critical angles for LEAN (α^{LEAN}) and PUSH (α^{PUSH}) for the same conditions. These figures show that increases in shoulder/wall friction increase PUSH force, while impeding the small LEAN forces. Note also that some LEAN and PUSH force may be generated, even when pushing on a frictionless surface.

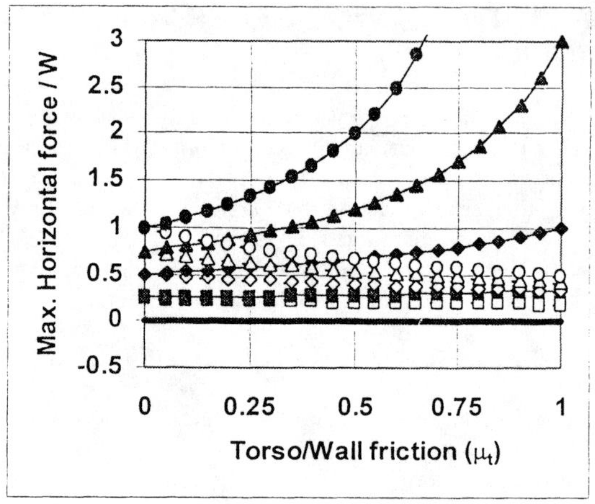

Figure 6. Horizontal forces generated by LEAN and PUSH normalized by body weight (W) as a function of μ_t for selected values of μ_b. No horizontal force can be generated when $\mu_b = 0$.

Critical Angles for $\bar{a}/L = 0.8$

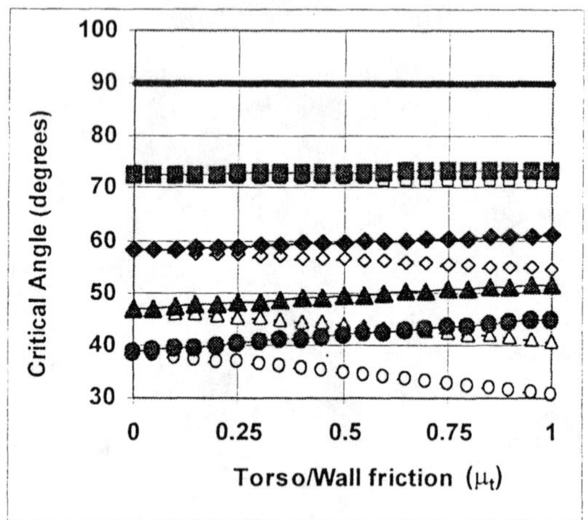

Figure 7. Critical slip angles in LEAN and PUSH as a function of μ_t for selected values of μ_b. Non-vertical orientations can only be achieved with non-zero feet/floor friction.

Legend for Figures 6 and 7:

F^{PUSH} Filled symbols F^{LEAN} Open Symbols

$\mu_b = 0.0$ Solid line
$\mu_b = 0.25$ Square icons
$\mu_b = 0.5$ Diamond icons
$\mu_b = 0.75$ Triangle icons
$\mu_b = 1.0$ Circle icons

The results governing PUSH force as a function of torso/wall and feet/floor friction are provided in the table below as these are of particular practical interest.

Table 1. PUSH forces normalized by body weight (F^{PUSH}/W) as a function of feet/floor and torso/wall friction coefficients.

Torso / Wall Friction (μ_t)	Feet Floor (μ_b) Friction				
	0	**0.25**	**0.5**	**0.75**	**1**
0	0.00	0.25	0.50	0.75	1.00
0.25	0.00	0.27	0.57	0.92	1.33
0.5	0.00	0.29	0.67	1.20	2.00
0.75	0.00	0.31	0.80	1.71	4.00
1	0.00	0.33	1.00	3.00	**Infinite**

CONCLUSIONS

Horizontal force generation (via either LEAN or PUSH) is not possible without footwear/floor friction. Indeed, little PUSH or LEAN is possible while standing on slippery surfaces (Figure 4). However, both LEAN and PUSH can generate horizontal forces in the absence of torso/wall friction (Figure 6). Examining equations 13 and 17, we note that $F^{PUSH} > F^{LEAN}$ for all combinations of wall and floor friction coefficients, except when μ_t=μ_b=0, in which case F^{PUSH}=F^{LEAN}=0, and when μ_t=0 . For a/L = 0.8, critical angles decrease with increasing footwear/floor friction (Figure 5), but they are fairly insensitive to torso/wall friction (Figure 7) for low values of foot/floor friction (μ_b < 0.5).

General observations on LEAN:

Horizontal force due to LEAN can never exceed body weight. LEAN forces become less for increasing torso/wall friction (μ_t), and increase with increasing foot/floor friction (μ_b). Critical angles during LEAN diminish (become more shallow) with increasing friction at the wall or floor.

General observations on generating PUSH:

Once some purchase of the feet is achieved, both increases in footwear/floor friction (μ_b) and in torso/wall friction (μ_t) increase the PUSH force that can be generated using axial thrust.

DISCUSSION

The mathematical results of this paper show that floor friction is the most critical factor in determining maximum PUSH capability. This finding is also demonstrated in military pushing specifications based on empirical data taken from MIL-STD-1472D (1989). Friction-related limits on achievable PUSH forces may exceed body weight for high-friction wall and floor conditions (Table 1).

The developments and results in this paper form an upper bound in determining LEAN and PUSH forces which cannot be exceeded, even given sufficient bodily strength to do so. This is demonstrated in Table 2, where empirical data obtained from MIL-STD-1472D (1989) are compared with the theoretical limits derived in this paper. In Table 2, theoretical values are given for a nominal wall/torso friction coefficient (μ_t) and empirical values of μ_b as given in MIL-STD-1472D. This allows a direct comparison between these empirically determined values and the theoretical values determined in our analysis. The theoretically determined limit is seen to be an upper bound for both the 5% and 95% male subjects. In all cases, the theoretical push limits exceed the empirically determined guidelines. The theoretical pushing limit for low friction conditions (μ_t = 0.2-0.3) is close to the guidelines. However, theoretical friction-related limits exceed the experimentally-based limits for increasing values of μ_b. This may indicate that the axial thrust required to generate the theoretically possible higher level of push is not achievable due to strength and ergonomic limitations (Chaffin *et al*, 1999)

Table 2. Horizontal push forces exertible intermittently for short periods of time.

Foot/ Floor Friction Coeff. (μ_b)	Max. Empir. Push kg (lbs)	Empirical F^{PUSH}/W for a 5% male 61.5 kg (135.8#)	Empir. F^{PUSH}/W for a 95% male 98.0 kg (216.3#)	Theoretical F^{PUSH}/W μ_t = 0.25 μ_b as given below
Low traction μ_b = 0.2 - 0.3	11.3 (25)	0.18	0.12	μ_b = 0.25 0.27
Medium traction μ_b = 0.6	20.4 (45)	0.33	0.21	μ_b = 0.6 0.71
High traction μ_b > 0.9	31.7 (70)	0.52	0.32	μ_b = 0.95 1.25

These findings may be summarized by the following observations. You can't push effectively while standing on ice, but you can push against an ice-covered wall. The effects of axial pushing on pushing or leaning force are subject to the limitations of the above relationships. Thus, two people of equal weight but of varying strength may be able to exert differing amounts of push on a given surface or object, but they are both subject to the same friction-related limits.

NOMENCLATURE AND DEFINITIONS

Wall	Vertical object subjected to the horizontal force generated by the leaning or pushing body
Floor	Horizontal surface supporting the leaning or pushing body
Torso	Portion of the body in contact with the Wall as defined above
Feet	Portion of the body in contact with the Floor as defined above
μ_b	Feet / Floor friction coefficient
μ_t	Torso / Wall friction coefficient
L	Body length
W	Body weight
$\bar{a}$	Effective length at which body weight acts
α	Angle at which incipient slip occurs at both footwear/floor and torso/wall interfaces
H_t	Horizontal reaction at the wall
H_b	Horizontal reaction force at the floor
V_t	Vertical reaction force at the wall
V_b	Vertical reaction force at the floor
F^{LEAN}	Maximum horizontal reaction force during passive LEAN
F^{PUSH}	Maximum horizontal reaction force during active PUSH
α^{LEAN}	Critical slip angle during passive LEAN
α^{PUSH}	Critical slip angle during maximum active PUSH

REFERENCES

[1] Military Standard MIL-STD-1472D. Human engineering design criteria for military systems, equipment and facilities, March 14, 1989.

[2] Chaffin, D. B., Andersson, G. B. J., and Martin, B. J., 1999, *Occupational Biomechanics, 3rd ed.*, John Wiley and Sons, New York.

[3] Barnett, R. L., 1996, "Ladder Slide Out - First Order Analysis", Triodyne Safety Brief, Vol. 12, No. 1.

Proceedings of
2001 ASME International Mechanical Engineering Congress and Exposition
November 11–16, 2001, New York, NY
SERA-Vol. 11

IMECE2001/SERA-24013

SITING AND SAFETY ASSOCIATED WITH A MUNICIPAL SKATE PARK

D. W. Pyatt, PE
Town Councilman
Town of Mount Airy, MD

ABSTRACT

The town of Mount Airy, MD chose to construct a skateboard and roller blade park in 1998 because of the many young adults skating on parking lots, streets, and sidewalks within the Town Limits. Many citizens and merchants complained of hazards both to the young adults as well as to pedestrians and drivers. Several sites were selected for evaluation; the final site selection was an existing Town Park. The next step was to design the Skate Park to satisfy both the recreational needs and safety of skaters of all ages and skill levels. In this capacity, the author worked with skating groups, including professional skaters, and insurance experts who insured the Town's facilities. A design was made, submitted for bids, and constructed in October 1999. The Skate Park has operated safely in the past year and has won recognition as a skater's park. This paper will discuss how this was accomplished.

INTRODUCTION

In the past few years, several companies began to produce skateboard ramps. A few companies were able to make composite skateboard parks. The Town wanted to build a skate park, but did not know precisely where to site the Park (or even what the criteria for siting should be). A rough budget estimate was made, but without an engineering study.

The Town Council approved this project with a budget of $100,000. Subsequently, Carroll County, MD contributed $32,000 to the project. The Council also found that few recognized safety standards existed. The Local Government Insurance Trust (LGIT) insures our facilities, and they were uncertain as to the additional premium cost for this facility.

SITE

The Town of Mount Airy has several parks; the largest one, Watkins Park, consists of about 50 acres. In addition, the Town has about 1,000 houses in the pipeline. We require new developments to provide 10% of the land as open space. Since visibility from main roads was considered to be highly desirable in order to minimize vandalism at night (subsequently this happened), we earmarked a new development and a small parcel of land as the site for the proposed skate park. It was reasonably close to a major road. However, nearby residents read about this in the local newspaper and complained at one of the monthly Council Meetings. We then decided to look at a new site.

We then identified Watkins Park, located adjacent to a state highway (Route 27), as the preferred site. The site suffered slightly, since the proposed location placed the Skate Park further from a main artery than desired. However, it was decided that this site had the best overall characteristics. In addition, it was owned by the Carroll County Board of Education, and the county provided about 30% of the funds.

DESIGNING AND BUILDING THE SKATE PARK

The design phase was next and perhaps the most challenging. Companies generally offered to do a turnkey project, with no firm cost estimates or specifications. Costs (verbal quotes) ranged up to $350,000. After a review of available information, we decided to design the skate ramps and put them out for bid. The first phase was selecting a 120-ft. square as the basis. Skateboard parks tend to be square or slightly rectangular in shape. After much research, we contacted a local skating store, and this proved to be the key. We found that the experts typically are teenagers, and professionals (those funded by skateboard manufacturers) are in their later teens. Several skateboarders sketched up their favorite designs on sheets. The ramps for beginners are smaller (lower) and spaced to keep speeds lower. For advanced skaters, the ramps are higher and allow for higher speed at ramp bottom. Since the author was a mechanical engineer, the sketches were made into engineering drawings (See Figure 1). The final design combined features of other skate parks and the input of our "experts."

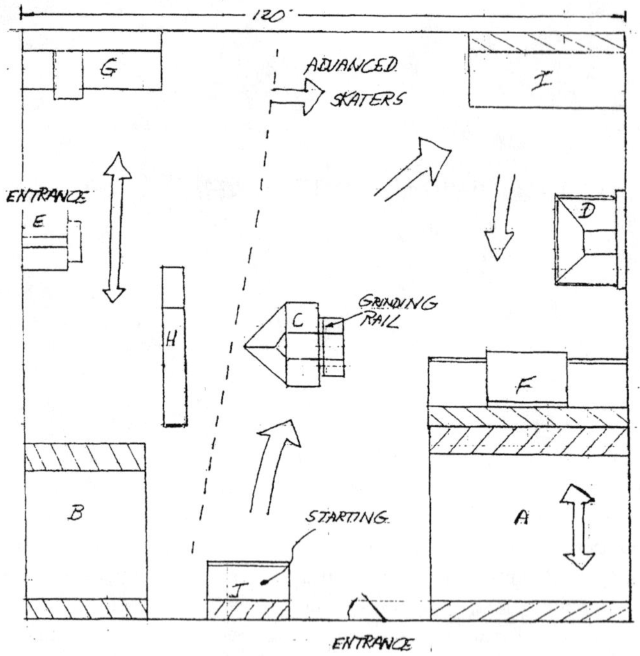

Figure 1. Engineering drawing of proposed skate park

Construction

The drawings and town specifications were prepared as a bid package. These packages were sent to about ten companies located throughout the U. S. A local company in West Virginia was selected that completed the project in October 1999. The completed park is shown in Figure 2.

Figure 2. Completed skate park

SAFETY CONSIDERATIONS

As mentioned, there are few recognized safety standards for skate park construction. Skaters are propelled roughly 15 ft. above the paved skating surface, and there is no doubt that serious injury or fatalities are possible. However, a research of injury records by the Consumer Product Safety Commission and referred to by the Local Government Insurance Trust from hospital emergency centers did not indicate a serious problem from skating injuries. This information also indicated:

- One-third of the victims had been skating for less than one week. Most of these were injured the first time they tried skating.
- Two out of every five injuries were sustained by a person using a borrowed skateboard.
- Those most frequently injured ranged in age from 10 to 14, and suffered 45% of the injuries.
- Fractures were the most common type of injury, occurring in about one-third of all injuries.
- One-half of all injuries were to the lower arm or leg.
- Five percent of the injured persons were admitted for hospital in-patient treatment.
- One out of every three accidents occurred when skaters struck irregularities in the riding surface. One out of every four accidents involved victims who lost their balance.
- Slightly over 1% of the injuries were attributed directly to the product.

In discussions with the skating design team, we discovered that they do get injured (typically broken bones in ankles and legs, ligament damage, contusions on knees and legs), but these are not often reported. There is a certain understanding that accidents are part of the sport; most of the legs and ankle injuries either heal in a reasonable time or prevent skaters from further injuries; i.e., forcing a "retirement." The skaters, though teens, are extremely dedicated to this sport, and just don't report many injuries, although these would be considered serious injuries to many other young men.

In a letter to the Town, LGIT stated:

"The three most important areas of risk management are supervision, protective gear, and rules/warning signs. Local governments planning skateboard and rollerblade facilities should consider supervision of the facility when it is open to the public. The person or supervisor in charge of the facility should have sufficient knowledge of the activity in order to determine the skill level of the skaters before allowing them to skate. The supervisor should also be trained in CPR and first-aid care. The skateboard facility area must be designed and constructed with a priority commitment to reducing the risk of injury to participants and spectators. An individual must check the equipment itself daily for any visible hazards or repair needs. A locked entrance should secure the designated skateboard area when the facility is closed. We also consider that you consider a hold-harmless agreement wherein the participants, parents, or guardians are required to sign a waiver, holding the Town harmless from bodily injury as a result of this activity."

LGIT also recommended that:

- Skateboard runs should be clearly labeled as to the degree of difficulty. (See Figure 3)
- The Facility supervisor should file a complete accident report on duty following any accident or injury occurring at the skating facility.

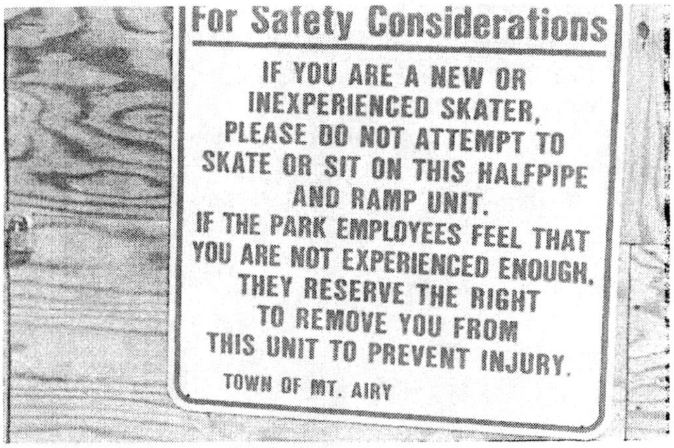

Figure 3. Safety sign posted near Skate Park

- Employees should be trained regarding all safety rules and procedures, operational procedures, and management requirements. All training and orientation given to skating area employees should be documented and retained on file.
- All users of the skating area should complete a registration card, which is shown to gain entrance to the facility. A waiver disclaiming the Town from injury claims should be signed (these are kept on record by the Town).
- A Facility Director should be appointed. His or her role would be to control the number of people using the facility, determine the skill level of skaters, and direct them to the proper ramps or runs. Employees should be certified in Red Cross First Aid, and an appropriate first-aid kit should be available on site. Appropriate contact with police and ambulance should be maintained.
- There should be a documented daily visual check by the employees on duty for any visible hazards or repair needs.
- A documented, complete inspection should be performed by the maintenance supervisor at least monthly; included in all inspections should be all walking or standing surfaces, fencing, steps, handrails, spectator areas, and any construction deficiencies.
- All maintenance and repair work should be documented as to the type of maintenance performed, the name of the employee performing the work, and the date completed.

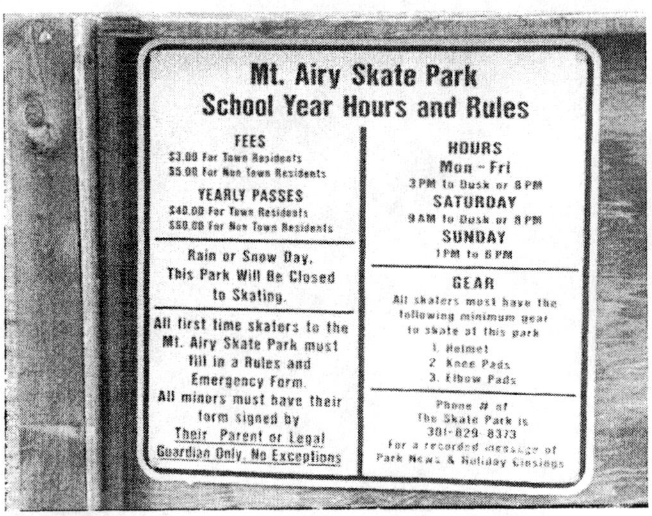

Figure 4. Safety gear requirements and rules posted

Safety Equipment

Safety equipment should be worn at all times while using the facility. This equipment shall include, but is not limited to, the following: an appropriate skateboard in good working condition, helmet, knee and elbow pads, gloves and wrist supports, and proper shoes. Additional equipment for consideration is specifically designed slip-resistant shoes, helmets, padded jackets, and padding for hips, knees, and elbows (see Figure 4). A safety fence was installed to prevent skaters from falling from ramps (see Figure 5).

Figure 5. Installed safety fence

As a result of the Town's active involvement in the design and operation of the skate park, including hiring and maintaining a staff of attendants, our annual insurance premiums – originally estimated to be $13,000 – was reduced to $3,500, saving the Town almost $10,000.

OPERATING EXPERIENCE

From a safety and goodwill standpoint, the park has been very successful. Only a few minor injuries have occurred. The skate park attendants are generally paired, one collecting the money and the second one directing the flow of skateboard and rollerblade patrons.

Since its opening in October 1999, the park has taken in about $75,000 in revenues. We have had several demonstrations by professional skateboard teams. The general consensus is that the park is very enjoyable, and is comparable to nationally ranked facilities, giving additional visibility to the Town.

ASME RESEARCH STUDY

Although in its infancy, the ASME Research Committee on Risk Technology has recommended that a project entitled *The*

Development of Risk-Based Standards for Skate Parks and Other Active Recreational Activities be funded by the Center for Technical Research.

CONCLUSIONS

A well-designed and operated skate park can be operated safely if it is properly supervised and maintained. The Town followed the guidelines suggested by our insurer, and maintained a good safety record.

REFERENCES

Guidelines for Public Skating Facilities (Skating/Rollerblading), published by Local Government Insurance Trust, March 31, 1999

Townsend, E.B., "Risk Management Guidelines for Public Skateboard and Roller Blade Facilities," letter to Mount Airy Mayor Gerald R. Johnson, July 2, 1999.

Proceedings of
2001 ASME International Mechanical Engineering Congress and Exposition
November 11–16, 2001, New York, NY
SERA-Vol. 11

IMECE2001/SERA-24014

THE EXPLOSION OF A FIREWORKS STORAGE FACILITY AND ITS CAUSES

B.J.M. Ale
National Institute for Health and the Environment
PO box 1, 3720 BA Bilthoven, The Netherlands

ABSTRACT

On 13 may 2000 an explosion occurred in a fireworks storage and trading facility in Enschede, the Netherlands. Twenty-two people were killed and some 900 injured. The material damage was approximately 400 MEuro.

Immediately after the accident an investigation was started into the causes of the accident. Special attention was given to the unexpected violence of the explosion. The investigative committee installed by the Government used results and advice of domestic and international institutes to obtain results.

It appeared that the firm had a long history of violating permits, that the city had legalised these violations and that inspectorates and state institutions were not aware of the hazards thus created. Especially the importance of the correct classification of the fireworks and of the storage of the correct types and quantities went unnoticed. As a result prior to May 13 2000 most of the fireworks stored at the premises were more powerful than the labels indicated and in fact a significant part of the storage was mass-explosive contrary to the current permit.

INTRODUCTION

On May 13, 2000, a series of explosions took place on the SE Fireworks premises in Enschede, The Netherlands, which destroyed the facility and caused considerable damage to the surrounding neighbourhood. The disaster caused the death of 22 persons, four of whom were fire-fighters. Approximately 950 persons were injured, may of whom seriously. The material damage is estimated to be over 500 MEcu.

Following the disaster in investigative committee was installed by the government. The task of this committee was to investigate the disaster, draw lessons for the future and in doing so help relief the suffering of the victims.

THE INVESTIGATION

The committee has followed two tracks in its investigation into the cause, the development and the extent of the fireworks disaster. These tracks have influenced each other in the course of the investigation

The first track was the development of the committee's own picture of the cause and the development of the disaster. This was done using an analytical framework, which comprised all potential hypotheses regarding the initiation and the development of the various fires and explosions and about the cause effect relationships. The suggestions and hypotheses brought to the attention of the committee by outsiders were also placed in this framework, as the committee deemed it of the utmost importance that all suggestions were addressed and none dismissed a priori.

The second track consisted of gathering and judging information that could be of importance to judge the validity of the hypotheses.

Important sources of information for the committee were the investigations performed by other organisations but the committee into the technical aspects of the disaster. In particular the criminal investigation has to be mentioned that was performed by and under the responsibility of the Almelo district attorney office. This technical investigation was primarily aimed at the cause, the development and the extent of the disaster. The institute for applied scientific research (TNO) and the Netherlands Forensic Institute (NFI) were involved in this investigation [NFI 2001].

The committee made arrangements with the DA's office and with the Twente police department about the exchange of information at the start of the inquiry. As a result the committee could have access to visual material important for its investigation. Furthermore information was exchanged on a regular basis.

In addition the committee used information that was assembled in the joint investigation of the state inspectorates. It could also use the premature report by a local consultant from Enschede, MSNP, which had contact with the local interest group of victims of the disaster. The safety experts of the committee had several discussions with MSNP.

The committee is of the opinion that the various investigations make it possible to make a reliable reconstruction of the disaster. The committee finds important support for its opinion in the fact that three reputable independent foreign organisations support its analysis namely the Health and Safety Executive of the United Kingdom, the Bundes Anstalt fur Materialprufung from Berlin, Germany and the Bureau for Alcohol, Tobacco and Firearms of the United States. As such there is broad agreement about the course of events at SE fireworks on May 13, 2000 and the explanation for the force of the explosion.

The committee published its findings on February 28, 2001. [COV 2001, COV 2001a]

In the following the main features are sketched of the course of events at SE fireworks on May 13, 2000 as it emerges from the reconstruction.

FIREWORKS

Fireworks are grouped according to classes corresponding to the possible effects from a fire or an explosion during transport, that is to say when the fireworks are packaged according to the regulations for transport. The classes are indicated by 1.1 to 1.4. The associated properties are presented in table 1.

Class 1.4 is usually associated with so-called consumer fireworks. Class 1.1 is associated with High Explosives. There is an important demarcation between the classes 1.4 and the heavier classes in that class 1.4 when combined with class 1.1 in a single storage compartment has never been found to have mass explosive properties, while the other classes will give rise to a mass explosion when combined with class 1.1 fireworks.

Table 1: Classes of fireworks and their properties

Class	Description
1.1	Explosives that have a mass explosion hazard. A mass explosion is one which affects almost the entire load instantaneously.
1.2	Eexplosives that have a projection hazard but not a mass explosion hazard.
1.3	Explosives that have a fire hazard and either a minor blast hazard or a minor projection hazard or, both but not a mass explosion hazard.
1.4	Explosives that present a minor explosion hazard. The explosive effects are largely confined to the package and no projection of fragments of appreciable size or range is to be expected. An external fire must not cause virtually instantaneous explosion of almost the entire contents of the package.

It has been known in some circles that fireworks are often classified as less hazardous than they really are. This reduces the shipping costs considerably. This was reported to public authorities after an explosion occurred in a fireworks manufacturing establishment in The Netherlands in 1991 [TNO 1991]. However no measures were taken to remedy this situation.

THE COMPANY

The business of SE Fireworks is assembling fireworks and setting them off at events. The company started at its present location in Enschede, The Netherlands, in 1978 with a limited storage of fireworks in a number of prefabricated concrete boxes of garages. In the course of the following 20 years SEF applied a number of times for a permit for extension of its capacity. All these permits were granted by the city council In the end SEF maintained a series of storages, bunkers and work areas. The work areas and a series of bunkers together form one structure. Around this a series of so-called MAVO boxes and shipping containers are positioned which all serve as storage area for fireworks. SE Fireworks was issued a permit under the environmental management act, which regulates the maximum load for each of the areas.

Construction

Bunkers or boxes for storage of explosives or fireworks must be constructed, among other things, in such a manner that the shock wave is diverted to a safe location should an explosion unexpectedly take place in the storage area. That means that the bunker or box remains intact with the exception of a pre-selected discharge opening. This discharge opening can be a door or the roof. The choice is partly decided by the location of the storage unit and the related location of a safe area. For example, when selecting a door as discharge opening, a dirt wall is often built facing the door so that the pressure wave and any possible pieces of rubble can be absorbed. (see also at "position of the storage units", below)

The bunker complex was originally designed for storage of 1650 kg of light fireworks in each box. As an alternative, 500 kg of strong fireworks were allowed for four boxes. Assuming that light fireworks contain nearly 30% active matter and strong fireworks 60% at maximum, the boxes therefore were constructed for about 500 kg TNT equivalent. (30% of 1650). Eventually there were 7000 kg of strong fireworks in the boxes. At 60% active matter this is 4200 kg TNT equivalent: eight times the amount for which the boxes were designed.

In 1979, storage of 2000 kg in MAVO boxes is approved. A firewall is not yet required at that time. In 1981 TNO deems a firewall with a minimum thickness of 20cm necessary and recommends allowing the wall to extend 50 cm beyond the MAVO box (thus in front, above and in back) [TNO 1981]. In 1997, a firewall of half-brick masonry between the MAVO boxes is required in the permit; this wall must extend 50 cm

towards the front and above the doors. The thickness of the wall is specified as half-brick masonry. In practice, as was discovered after the disaster, these walls were never erected.
The fire resistance of the different storage areas is intended to prevent escalation of a fire from occurring so quickly that repressive action is no longer possible.
The fire resistance of the MAVO boxes is insufficient, as shown in a report from TNO. In that report erecting firewalls between the boxes is therefore recommended.
In a letter from the army to Enschede in 1994 fire resistance is required for the containers of 120 min. and for the container doors of 30 min. In the 1997 permit, the wooden doors of the containers are required to have a fire resistance of 30 min. However unmodified ISO-containers have a fire resistancy of 4 minutes only [TNO 1997].
In the July 1999 permit no requirements are set for the fire resistance of the shipping containers. Building firewalls as for the MAVO boxes was not considered. It is even questionable whether this would help. The walls are still not fire resistant in that case. Furthermore, fire resistance does not work either in this case against sympathetic reaction jumps. According to the letter from TNO mentioned above a standard 20-foot container has a fire resistance of only 4 minutes.
No indications have been found (until now) that the shipping containers which were being used on the SE Fireworks premises for storage of fireworks had in any way whatsoever been adapted to improve fire resistance.

Position of the storage units

It is important for what follows below that when there is first mention of storage of fireworks on the SEF premises, 7 MAVO boxes INSIDE a warehouse are referred to. This temporary provision was already approved in 1976 in a declaration of no objection. In 1977 the declaration of no objection was issued once again. In 1978 construction of the bunker complex was initiated, but this was not ready in time for the 1978 end-of-the-year campaign. Once again a declaration of no objection is issued for the use of the MAVO boxes. In 1979 SEF requests permission for continued use of the MAVO boxes, among other things. Anticipating the permit to be granted, another declaration of no objection is issued in August 1979. On December 14 1979 the permit is granted. In 1989 a container is permitted for "temporary storage" of fireworks for the first time.
The sate advisory unit for the storage of explosives, MILAN, recommends in a letter for the 1979 permit [MILAN 1997] dated August 27, 1979, to alter the position of the MAVO boxes to prevent the doors of these boxes to face the doors of the bunkers. This recommendation was not followed. A similar recommendation is made in 1994 regarding the positioning of the containers in such a manner that they do not face any other storage units [MILAN 1994]. That recommendation was not followed either. In the end the doors of several of the MAVO-boxes and several of the containers face the doors of the central bunker complex.

Table 2: Allowed quantities

Room	Allowed quantity (gross)
Big bunker central complex (C3-C11)	7.000 kg 1.4 S/G
Bunker C13 central complex	7.000 kg 1.4 S/G or 500 kg 1.3 G
Small bunkers central complex (C12, C14 and C15)	5.000 kg 1.4 S/G or 500 kg 1.3 G
MAVO-boxes M1 – M7	3.500 kg 1.4 S/G
Containers E1 – E14	3.500 kg 1.4 S/G
Total	158.500 kg 1.4 S/G or 136.500 kg 1.4 S/G and 2.000 kg 1.3 G

In the description of the works and of the events below the text often refers to number codes of storage units. The various storage units and their numbering is represented in figure 1
The maximum load as presented in table 2. In total, SE Fireworks has permits for 158,500 kg class 1.4 fireworks and 100 kg class 1.1 fireworks. As an alternative, 2000 kg class 1.3 fireworks can be stored, but in that case the maximum amount of class 1.4 fireworks is 136,500 kg.

The situation of room C2 on May 13, 2000.

The process of escalation that lead to the disaster started in the workroom C2 of the central compartment (the repackaging room), where the fire started and the first fireworks exploded. The events in room C2 deserve special attention. None of the organisations involved have been able to answer the question of the cause of the first fire in C2 in such a way that a definitive and unique statement can be made. Room C2 also deserves special attention because this room was the only room according to the temporary permit of change of July 19, 1999, in which the presence of unpacked fireworks to a maximum of 500 kg 1.4S of 1.4G was allowed. This made the situation in C2 especially vulnerable to safety problems compared to the other compartments. The presence of unpacked fireworks only was allowed during working hours.
The construction of C2 consisted of a reinforced concrete skeleton of walls and a roof. In the roof were 6 light cupolas. There was an access door to C2 at both sides of the central

storage building. No automatic devices for the detection or reporting of a fire were present. The central storage building was fitted with a "sprinkler system" in the form of hand operated spray heads, the lines of which ran through C2. Presumably there was no water in the lines of the installation and thus it was not operable. In C2 there was also an improvised electricity line, constructed by workers of SE Fireworks themselves. From statements made by people involved it can be concluded that fireworks were present in C2 permanently - thus on may 13, 2000 also - but that no work was done on may 13, 2000.

On may 13, 2000 approximately 900 kg fireworks was present in C2, among which cake-boxes, rockets and 55 to 60 shells of various calibre. Fireworks have been present in C2 over longer periods of time and works was being done on it. This generally leads to the presence of pyrotechnical dust that accumulates and spreads all over the room.

Fireworks on the premises on May 13, 2000

Obviously it was not possible to determine the amount and nature of the fireworks present at the premises prior ro the explosion, as the works were completely demolished by the accident. However, containers en route to SE Fireworks were seized by the authorities and analyzed for their classification. Also statements by witnesses were used to determine the distribution of the fireworks over the compartments.[TNO 2001, BAM 2001]

From the analysis it appeared that of the 177 tons of fireworks present at the location 16 tons were class 1.4, 153 tons were class 1.3, 5 tons were class 1.2 and 1.7 tons were class 1.1. This brings the amount of fireworks legally present to approximately 10% of the total quantity. It should be noted that according to statements made by the owners all fireworks were labeled 1.4

In addition there were several compartments that presumably had mass explosive properties due to the fireworks stored in them. These compartments are listed in tabel 3.

Table 3: Mass explosive compartments om May 13, 2000

Compartment	Type fireworks 1.1	Other firworks	Quantity (kg)
Minimal two statements			
C2	Unpacked shells	Rockets, cake boxes, fountains	900
C9	Titanium salutes	Shells 3", 4"	7.344
E2	Shells 12"	Shells ≥ 5"	3.917
One statement			
C11	Titanium salutes	Shells 3"- 8 "	4.896
C15	Shells 12"	Shells ≥ 8"	2.709
M7	Shells 12"	Shells ≥ 6"	2.447
E8-E11	Titanium salutes	Shells 3"- 10", cake boxes	19.583
TOTAL			41.796

THE COURSE OF EVENTS ON MAY 13, 2000

Around 15:03h a first fire alarm reaches the alarm central of the regional fire departement. The first unit arrives at the scene at 15:08. The site is closed and nobody is present.

The fire-crew determines that room C2 of the central storage building is on fire and that parts of the front of this room are on the ground. As far as the doors of the other compartments are visible, they are closed. There are additional fires at the front of building H, on the roof of building G and on the roof of several of the MAVO boxes. There are also fires on several places outside the establishment, among other in the shrubbery in the Tollensstreet at about 60 m away from the storage facility, probably caused by firework items thrown from the fire. The fire department assumes that they have conquered the fire in room C2 at about 15:15.

After opening room C3 at about 15:18 and having drenched the fireworks in it as a precaution, the fire departement detects fire in room C4. Behind this room fireworks explode. Also fireworks are launched from C4. These fireworks explode in the surroundings causing additional fires. At about 15:28 the fire department sees smoke emerging from between containers E2 and E15. Two minutes later a high-pressure hose is directed to the fire from the roof of container E1 to fight the fire between the containers E2 and E15. At about 15:32 a low-pressure jet is directed from the roof of MAVO-box M6 to room C4. It looks like the intensity of the fire there is decreasing.

Then, at about 15:34, several members of the fire crew hear a loud rumbling noise. From containers E2-E15 a jet flame shoots over the ground in the direction of the central storage. Container E2 expands and an explosion takes place by which the roof for a moment bends upward. After another 18 seconds container E2 explodes by springing its doors open. A lot of fireworks are expelled from E2 among which shells. The effect-charge of these shells explodes where the shells land. Some 42 seconds after the explosion of E2 the whole row of MAVO-boxes explode, starting with M7 at about 15:35. The explosion of M7 was the most powerful of this series.

The force of the explosion of the MAVO-boxes forces the doors of the compartments of the central storage inward. About 66 seconds after the explosion of the MAVO-boxes the whole of the central storage explodes starting with room C11. The remaining containers also explode.

THE CAUSES

In report A part III the committee has given its own analyses and judgement of the course of events on May 13, 2000. Next follow the main lines of this analysis and this judgement.

The first fire in C2.

Statements by various witnesses indicate that fireworks have been exploding at SE Fireworks from approximately 15 minutes before the first alarm reached the alarm central of the regional fire department (which happened at about 15:03). A number of witnesses had the impression that fireworks were set off deliberately. However there aren't any witnesses who have seen with their own eyes that fireworks were set off on the premises if SE fireworks on may 13, 2000. Therefor it is also possible that it concerned fireworks that was ignited by the fire in C2 and was launched through the destroyed light cupolas. From what is known by the committee about the fireworks that were present at SE Fireworks on May 13, 2000 it appears that the fireworks present in C2 could indeed have caused this effect. The three firemen interviewed by the committee also heard rumbling noises before 15:03. As soon as the fire department arrived at the scene the discovered at the first superficial exploration of the premises that there was a fire in C2 and that parts of the front were lying on the ground in front of the compartment.

Possible causes of the fire in C2

In report A part III all thinkable causes of the fire in C2 have been described. Some of these causes could be dismissed immediately for various reasons. The remaining possible causes are the following.

A. Human action

1 Human action that was aimed at causing the fire and that had

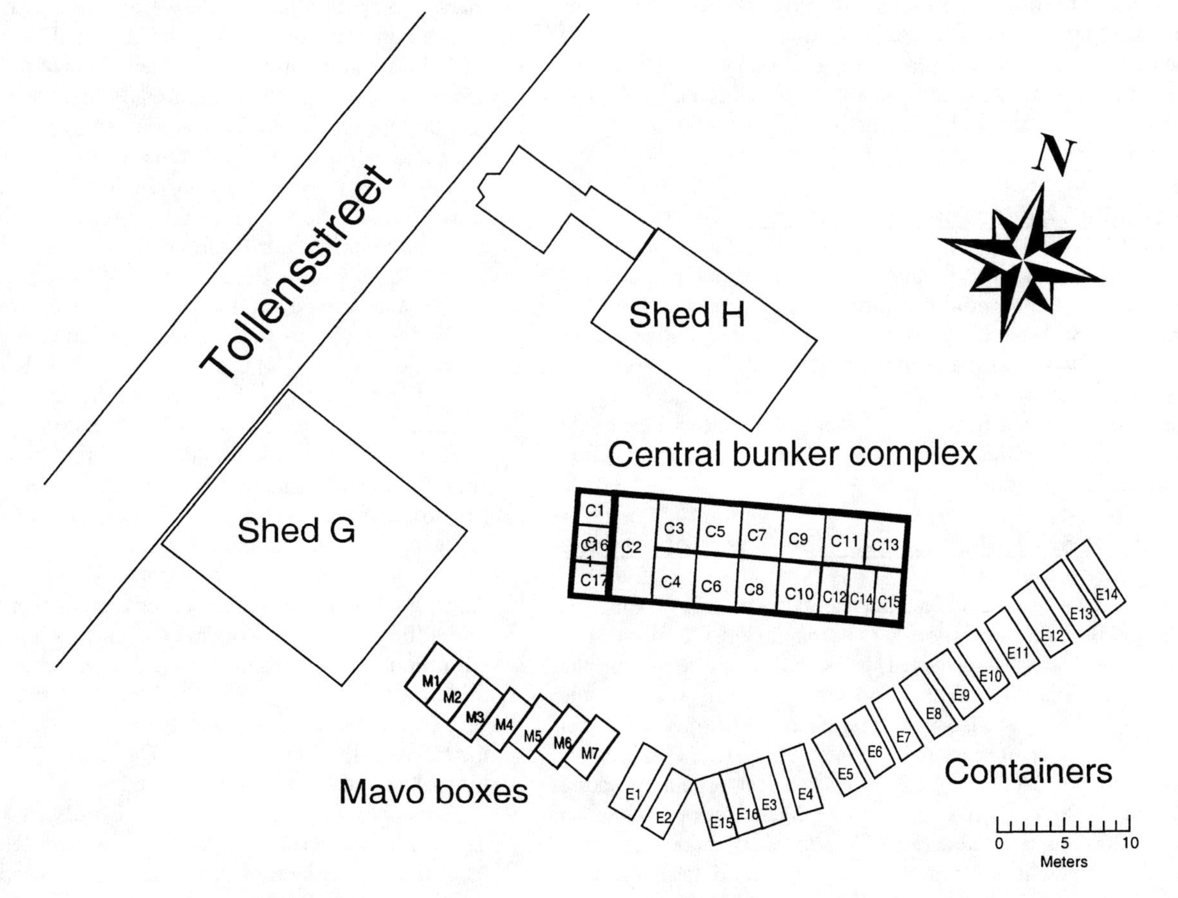

Figure 1: Layout of the SEF works.

the fire as approximately immediate effect: arson or sabotage.

2 Human action that was not aimed at causing the fire but lead to the fire anyway as a more indirect effect.

a: Clandestine setting off fireworks at an earlier time causing the fire in C2 as an unintended consequence, or as the result of an incident during the setting off.

b: Negligence at the last operations performed in C2 resulting in (auto-)ignition of the fireworks in C2, before or after the people concerned had left.

B: Technical factors

1 Autoignition of flashcord present in C2, with or without the combination with the pyrotechnical dust accumulated and spread in C2 over time, or of other material sensitive to (auto-)ignition or explosion.

2 Short-circuit in the electrical system in C2.

Ad A1: Arson or sabotage.

On January 26, 2001 a person is taken into custody on suspicion of being involved in the starting of the fire in SE fireworks. This person is still held in custody to date. This makes arson still a potential cause of the fire in C2 with more than minimal likelihood. The further results of the investigation have to be awaited. Only in case of a confession, and or a conviction at which arson in C2 is proven, arson can be designated as the cause of the fire in C2. Only than all other remaining causes can be dismissed. Until such time arson remains one of several possible causes.

Ad A2a Clandestine setting off of fireworks.

As was indicated above, a number of witnesses have stated that they had the impression that fireworks were set off on the premises of Se Fireworks before 15:00. The committee deems setting of fireworks as cause of the first fire in C2 less probable. None of the witnesses have actually seen the fireworks being set off and the observed fireworks as was mentioned earlier, could also have been initiated by the fire in C2 and subsequently thrown from this room.

Ad A2b Negligence at the last operations performed

The committee also deems negligence at operations on may 13, 2000 less probable as cause of the first fire in C2, because it judges it unlikely that indeed work has been done on the premises on may 13, 2000. The fire department found the premises locked and also interrogation by the police have not lead to indications of work being done on may 13, 2000.

The committee deems negligence on the last regular working day, May 12 2000, more likely than negligence at operations on may 13, 2000 mentioned earlier. Also on this point no definitive statement can be made. On May 13, 2000 dried flashcord was present in C2; flashcord was dried in C2 on a wooden pole over the central heating. In certain circumstances (e.g. when in contact with pyrotechnical dust) flashcord is sensitive to auto-ignition. It therefor cannot be excluded that auto-ignition of the flashcord in C2 has taken place, which subsequently may have lead to the ignition of the fireworks present in C2.

Ad B1 Auto-ignition of the flashcord in C2

The possibility of auto-ignition of the flashcord in C2 was already mentioned above. The committee cannot exclude this possibility. This holds especially when potential negligence at works, the situation in the room as such and the heating of C2 as result of the weather conditions on May 13, 2000 are taken into account.

Ad B2 Short-circuit.

The committee also cannot rule out short-circuit as possible cause for the first fire in C2. The main reason is the presence in C2 of improvised electrical wiring, constructed by employees of SE fireworks themselves.

The further escalation of the fire in room C2 to room C4

The fire in C2 only was noted after external effects had taken place, in particular ignited fireworks that landed in the surroundings while burning and the rumbling noises that were audible at large distances. When the alerted fire department arrived at the scene, the fire already had 25 minutes to develop and the ejected fireworks had caused fires at various places in the surroundings at large distances from C2. At a certain point in time fire also began in the triangle between the containers E2 and E15 and the concrete fencing around the site. This fire could have started fairly early.

Even after the fire department had mastered the fire in C2 smoke and later even flames detected emerging from C4. It cannot be excluded that fireworks emitted from C2 penetrated the door of C4 and caused the fire in C4. However it is more likely that the fire travelled through the opening of the sprinkler system. The report by NFI/TNO mentions a hole in the wall between C2 and C4 for the waterline with a diameter of approximately 7 cm. Several mechanisms for the travel of the fire are thinkable such as:

- direct travel through the hole
- development of heat in the (probably empty) copper water-line causing the melting of the solder resulting in hot pieces of the line falling on the fireworks in C4 causing it to be initiated
- Ignition of pyrotechnical dust that accumulated over time on this water line.

The committee cannot dismiss any of these potential causes of the travel of the fire from C2 to C4. There could have been even more than one mechanism involved.

Subsequently the fire department also discovered the fire in the triangle between E2, E15 and the concrete fencing. As appears from the reconstruction of the actions of the fire department as given in report B chapter 4, the fire department has tried to extinguish this fire. The detection of this fire was difficult because there was no direct access to this triangle. The fire could only be extinguished from above, in particular from E1. In the triangle stood among other an old trailer.

The explosion of container E2 and the resulting further escalation

It is plausible that the fire in the triangle already had progressed to the contents of container E2 when the fire department tried to extinguish it. This can be concluded among other from the fact that shortly after the fire department started extinguishing activities aimed at the triangle, fire was discovered in container E2 which exploded shortly afterwards.
The limited fire-resistance of the standard container can be named as the direct explanation for the rapid transfer of the fire to the contents of container E2. TNO estimates the fire-resistance of a standard container to be 4 minutes. The speed at which the fireworks in E2 were initiated is presumably also. Circumstances that also played a part in the fire and the explosion of container E2 were the absence of fire detection and alarm measures and the absence of a sprinkler system in the containers, the absence of a well sized relief opening and the presence of heavy fireworks in containers E2 (1.3 and possibly 1.1 fireworks).
The explosion of E2 resulted in a large jet flame that impinged on the central storage building. According to the analysis of NFI/TNO however further escalation was caused by the heavy fireworks expelled from the container. These fireworks exploded among other in from of the MAVO-boxes and give its explosive power could knock through the doors of these boxes and initiate the contents. The committee notes that initiation of the MAVO-boxed also could have happened via E1.

The explosion of MAVO-box M7 and the other MAVO-boxes

Within one minute after the explosion of container E2 the MAVO-boxes exploded in rapid succession starting with M7. The committee designates the strength of the fireworks as the primary cause of the explosion in combination with minimal damping between the boxes.
The fireworks present in M7 probably have to be considered as type 1.1. From the investigation of NFI/TNO it has to be concluded that it is improbable that the fire retardant walls that were prescribed actually were present. These walls could not have prevented the explosion to progress from M7 to the other MAVO-boxes anyway. For that a more substantial damping is necessary, for example by using a heavier construction or, preferably, more distance between the boxes.
The MAVO-boxes were not fitted with fire-detection, fire alarm of fire extinguishing devices either. These probably would not have had any effect anyway in view of the way in which the fireworks in M7 were initiated and the mass-explosion that followed immediately. The shock wave resulting from this explosion is estimated to have been equivalent to that of 800 kg of TNT. The fireball that went with this explosion had a diameter of 85m.

A little over a minute after the explosion of the MAVO-boxes the last explosion took place, in which C11 and subsequently the remainder if the central storage building were destroyed. Almost simultaneously also the remaining containers exploded.
It sufficiently probable that ejected fireworks and or the fireball from the explosion of the MAVO-boxes initiated the explosion in C11 because the doors of the central building already had disappeared as a result of the force of the explosion of the whole row of MAVO-boxes. The remaining compartments of the central storage building exploded either by initiation by fireworks expelled from the MAVO-boxes or other storages or from the heat of the explosion in C11. The time between the series of explosions is so short that it is unlikely that these are perceived as more than a single explosion.
The last explosion had the character of a mass explosion. The centre of the explosion lay in C11. In this bunker again heavy fireworks were probably stored (among which potentially 1.1 fireworks). Some of the containers reacted immediately. From the reconstructed contents it is probable that these contained heavy fireworks also.
The explosion had enormous power. The committee agrees with the estimate made by NFI/TNO on the basis of the resulting damage of the strength of the explosion to be equivalent to approximately 4-5 tons of TNT.
The committee explains the force of the explosions primarily from the strength of the fireworks present, as follows from the reconstruction, in combination with insufficient damping between the storages. The committee notes that the central storage building was constructed adequately for accidents with 1.4 fireworks but insufficient for the heavy explosions as took place in C11.

CONCLUSIONS

The committee concludes that the course of the disaster, in which a fire that started in a single compartment ultimately resulted in a fatal mass explosion of the central storage building, can be explained by two factors. Firstly that according to the reconstruction of the committee at may 13, 2000 on the premises of SE Fireworks 177000 kg fireworks were present of which probably more than 90% should have been classified heavier than 1.4. Secondly the way these fireworks were stored, namely in part as a combination of 1.3 and 1.1 fireworks. The committee points out that according to the license 158500 kg of 1.4 fireworks or 136500 kg 1.4 fireworks and 2000 kg 1.3 fireworks were allowed. If only these fireworks, with the right classification, would have been present a disaster such as the present one could not have happened. This is confirmed by a risk analysis performed by RIVM [RIVM 20001] at the request of the committee and by Maximal Credible Event analyses performed by TNO. [NFI 2001, loc cit]
The responsibility lies according to the committee with SE Fireworks. The State is to blame for not acting on the problem of the incorrect classification.
Only a limited number of factors remain as potential causes fir the explosion, of which arson is not the least probable. Also of

importance is that in C2 insufficient measures were taken for the safe handling of flammable and explosive materials. As a result auto-ignition of flashcord that was dried in C2 and short circuit in an improvised electrical wiring cannot be excluded as potential causes.

The explosion of fireworks following the initial fire was possible because SE Fireworks did not adhere to the existing permit. According to the permit no fireworks were allowed to be present in C2 outside working hours. No work was done on May 13, 2000 but nevertheless 900 kg of fireworks were present in C2.

The initial fire in C2 could evolve and escalate because there was no automatic, functional fire detection, fire alarm and fire extinguishing system. Because the improper way the "sprinkler system" was constructed the fire resistance between C2 and C4 was seriously deteriorated. It should be noted that, without reducing SE Fireworks primary responsibility, the permits were insufficient on this point. Would the permit have been adequate and would the permit be maintained, the initial fire could not have had such devastating consequences.

Container E15 has played a major role in the escalation of the fire. By its position in combination with E2, the triangle between E2, E15 and the concrete fencing was hardly visible and not accessible. In addition combustible material was present in this triangle such as the abandoned trailer that has been there, illegally, already looming before the disaster.

LITTERATURE

BAM 2001	Prufbericht, Berlin 15 December.
COV 2001	De Vuurwerkramp, Eindrapport, Commissie Onderzoek Vuurwerkramp, Enschede/Den Haag, Februari 28, ISBN 90-71082-67-9.
COV 2001a	De Vuurwerkramp deel A, SE Fireworks, De overheid, De ramp, Commissie Onderzoek Vuurwerkramp,/ Enschede/Den Haag, Februari 2001, ISBN 90-71082-68-7.
MILAN 1979	Letter HW/OS/38.631 August 27, Ministry of Defence, Den Haag..
MILAN 1994	Letter 27981/B October 5, Ministry of Defence, Den Haag
NFI 2001	NFI/TNO Onderzoek naar aanleiding van brand en explosies bij SE Fireworks op 13 mei 2000, 1 februari, Ministerie van Justitie, den Haag.
RIVM 2001	P.A.M. Uijt de Haag, G.M.H. Lahey, C.J.W. Twenhofel, B.J.M. Ale, Risico-analyse S.E. Fireworks te Enschede, RIVM, repnr 026/2001 projectnr M/408130/01/AH, January
TNO 1981	A. Harmanny, De opslag van vuurwerk in MAVO boxen, TNO, Delft, Maart 1981.
TNO 1991	W.P.M. Mercx, Verslag van het Prins Maurits Laboratorium TNO betreffende de explosie van de vuurwerkfabriek "MS Fireworks" te Culemborg op donderdag 14 februari 1991, TNO 292191122, TNO Delft.
TNO 1997	J.C.A. van de Weijgert, P.W. van der Haar, Rapport in briefvorm over de brandwerendheid van zeecontainers door TNO bouw in opdracht van het Arrondissementsparket te Maastricht, 7 oktober.
TNO 2001	E.G. de Jong, M.W.L. Dirkse, Classificatie onderzoek evenementen vuurwerk SE Fireworks deel2: Classificatie experimenten. Report nr PML 2001-C29, February TNO Delft.

Proceedings of
2001 ASME International Mechanical Engineering Congress and Exposition
November 11–16, 2001, New York, NY
SERA-Vol. 11

IMECE2001/SERA-24015

RISK ANALYSIS OF THE STORAGE FACILITY OF S.E. FIREWORKS IN ENSCHEDE, THE NETHERLANDS

P.A.M. Uijt de Haag, G.M.H. Laheij, B.J.M. Ale
National Institute of Public Health and the Environment
P.O Box 1, 3720 BA, Bilthoven,
The Netherlands
Tel.: +31-302743713, Fax: +31-302744428,
e-mail: paul.uijt.de.haag@rivm.nl

ABSTRACT

In the course of the investigation into an explosion in a fireworks storage facility in the Netherlands risk analyses were performed on the facility as it should have been under the current legislation and permits and as it actually was. The results show that under the current legislation the facility was acceptable in its present location, if it adhered to the terms of its permit. However as it actually was it was completely unacceptable.

The results are compared with the results of analyses of industrial facilities and other hazardous activities to place the risk of storage of pyrotechnics in the context of other industrial risk.

INTRODUCTION

On 13 May, 2000 a fire occurred at a storage facility of fireworks, S.E. Fireworks, located in the middle of the city of Enschede, the Netherlands. The fire resulted in a number of consecutive explosions. The final explosion had an equivalent of approximately five tonnes of TNT and resulted in the death of 22 persons, firemen as well as bystanders (Committee to Investigate the Fireworks Disaster, 2001, Ale, 2001). At the time of the explosion, people were mostly outdoors due to the good weather. Otherwise, the number of deaths might have been larger.

The storage facility S.E. Fireworks was located in a densely populated area. However, a risk analysis of the storage facility was never made, presumably because the storage was intended for class 1.4 fireworks and no risks for persons living in the surroundings were associated with this class of fireworks. To determine the risk to the people in neighbouring houses, a risk analysis of the storage facility is made based on the knowledge after the event. The risk analysis is carried out for two different situations, namely for the situation licensed, where the amounts of fireworks of classes 1.3 and 1.4 (ADR) stored are in accordance with the amounts licensed, and for the actual situation, where part of the fireworks is in reality class 1.1.

This paper describes the results of the risk analysis. First, the criteria in the Netherlands concerning the external safety are given, i.e. the risk of an establishment to the persons living in the surroundings of the establishment. Next, the situation at the storage facility of fireworks is described. Next, the risk analyses of the storage facility is described, based on the situation licensed and on the actual situation. Finally, the results of risk analyses are presented and compared to the regulations and other establishments.

CRITERIA FOR EXTERNAL SAFETY

The regulations for the transport, use and storage of dangerous substances in the Netherlands are based on the risks to persons living in the surroundings of the activity. To evaluate the acceptability of the activity with dangerous substances, two risk criteria are developed, namely the individual risk and the societal risk (Bottelberghs, 2000).

The individual risk represents the probability of an individual suffering death due to the activity with dangerous substances. The individual is assumed to be unprotected and to be present permanently, i.e. 24 hours a day and 365 days a year. The individual risk is presented as contour lines on a topographic map. In the case of a new situation, e.g. the siting of a new establishment, the extension of an existing establishment or new housing developments, the maximum allowable individual risk is equal to 1×10^{-6} per year; inside the individual risk contour of 1×10^{-6} per year dwellings are not allowed. For existing situations, i.e. houses already present near an existing establishment, the maximum allowable individual risk is equal to 1×10^{-5} per year.

The societal risk represents the probability of having an accident with N or more people being killed simultaneously. In the calculation, the people involved are assumed to have some means of protection, for instance by staying indoors. The societal risk is intended as a measure to indicate the disruption of the society after the accident and is presented as an FN curve, where N is the number of deaths and F the cumulative probability of accidents with N or more deaths. For the societal risk of an establishment an indicative limit is defined (Minister of Housing, Spatial Planning and the Environment, 1994), given as the limiting probability (F_{lim}) for the occurrence of an event with N or more deaths (Eq. 1):

$$F_{\lim}(N) = \frac{10^{-3}}{N^2} \quad (1)$$

The relationship begins at N = 10 and F = 1 × 10^{-5} per year.

DESCRIPTION OF THE STORAGE FACILITY

The storage facility of S.E. Fireworks was situated in a densely populated area in the middle of the city of Enschede, the Netherlands. The facility consisted of 36 separate storage units accessible from outside, namely bunkers (C3 to C15), MAVO boxes (M1 to M7) and containers (E1 to E16). One unit, C2, was used for the composition and repair of fireworks. The layout of the storage facility area is shown in Fig. 1. The licensed amount of fireworks and the distribution over the various storage units is listed in Table 1.

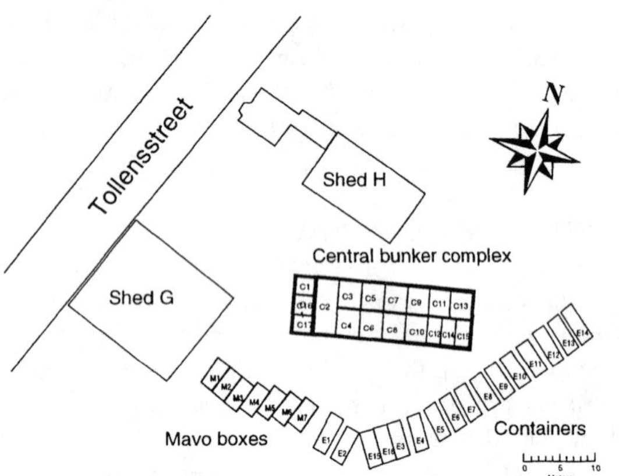

Figure 1: Layout of the storage

Table 1: Amount of fireworks licensed in the various depots

Depot	Class 1.4 fireworks		Class 1.3 fireworks
C2	500 kg*		
C3 to C11	7000 kg		
C12, C14, C15	5000 kg	or	500 kg
C13	7000 kg	or	500 kg
MAVO box	3500 kg		
Container	3500 kg		

*during working-hours

The classification of the fireworks is carried out by the manufacturer and, in the Netherlands, the classification is not verified by the authorities. It appeared that the classification of the fireworks present at the storage facility was not correct; part of the fireworks was classified as 1.4 but had the properties of class 1.1 explosives and was mass-explosive (Committee to Investigate the Fireworks Disaster, 2001). Thus it could happen that class 1.1 explosives were present at the storage facility although the license did not allow for it.

RISK ANALYSIS OF THE STORAGE FACILITY

To determine the risk of the storage of fireworks at S.E. Fireworks, two different situations are reviewed. In situation 1, the situation licensed, the establishment keeps strictly to the limitations of the licence, and the amounts of fireworks stored and the classes of fireworks stored agree with the amounts and classes described in Table 1. In situation 2, the actual situation, the amounts of fireworks stored agree with the amounts licensed. However, although the fireworks are labelled class 1.4, in reality part of the fireworks has the properties of class 1.1 explosives and is therefore mass-explosive.

The risk analysis consists of a number of consecutive steps. First, the probability of an accidental fire is determined. Next, the consequences to humans are calculated, given the accident. Finally, the probability of an accidental fire and the corresponding consequences are combined into the individual risk contours and the societal risk curve. In the analysis it is taken into account that the storage facility is situated in a built-up area.

SITUATION 1 – THE SITUATION LICENSED

The probability of an accidental fire

The accident which may end in effects outside the boundary of the establishment is fire in one of the storage units. In total, there are 36 storage units and all storage units are accessible from outside. According to the risk analysis method, each storage unit should therefore be considered as a separate storage unit in the determination of the probability of occurrence of a fire (Minister of Housing, Spatial Planning and the Environment, 1997).

The probability of fire in a storage unit depends among others on the safety procedures maintained at the establishment. The probability of an explosion in a (military) storage facility of explosives is estimated to be 1 × 10^{-5} per year (CPR, 1999). However, at a military site the safety procedures are probably more strictly than at a storage facility of fireworks and therefore the probability of an explosion at a storage facility of fireworks is likely to be higher. A storage facility of fireworks is presumably comparable to a warehouse of chemicals. The probability of a fire in a warehouse of chemicals is assumed to be equal to 8.8 × 10^{-4} per year (Minister of Housing, Spatial Planning and the Environment, 1997). In the risk analysis, the probability of a fire in a storage unit is therefore set equal to 8.8

$\times$ 10^{-4} per year per storage unit. Since the storage facility consists of 36 storage units, the probability of fire in one of the storage units is equal to 0.03 per year.

Consequences of the accident

The fireworks stored are mainly class 1.4 fireworks. Based on the characteristics of class 1.4 fireworks, a fire in a storage unit would lead to a flame emerging from the door of the storage unit with a length of maximally 15 metres (Handbook of environmental licenses). The length of 15 metres is an upper limit; in practical situations the flame lengths are restricted to a few metres distance from the door of the storage unit (TNO, 2000). Along with the class 1.4 fireworks, the license allowed the storage of small amounts of class 1.3 fireworks. Since the amounts of class 1.3 fireworks stored were small and storage was only allowed in the original packing, the effects of the class 1.3 fireworks are also restricted to a flame emerging from the door of the storage unit.

In the calculation of the effects on humans it is assumed that within the flame, the probability of death is equal to one. The heat radiation of the flame is assumed to be relatively low and therefore no lethal effects outside the flame envelope are expected.

Risk of the storage facility

The probability of a flame is equal to 8.8 $\times$ 10^{-4} per year per storage unit. The probability of death within the flame is equal to one and outside the flame equal to zero. Hence the individual risk contours of 10^{-5} per year and 10^{-6} per year coincide and are given by the envelope of the flames from the various storage units. An example of a flame from one storage unit and the individual risk contours of 10^{-5} per year and 10^{-6} per year are shown in Fig. 2. Within a distance of 15 metres of the storage facility no residents live. Therefore, no societal risk is calculated.

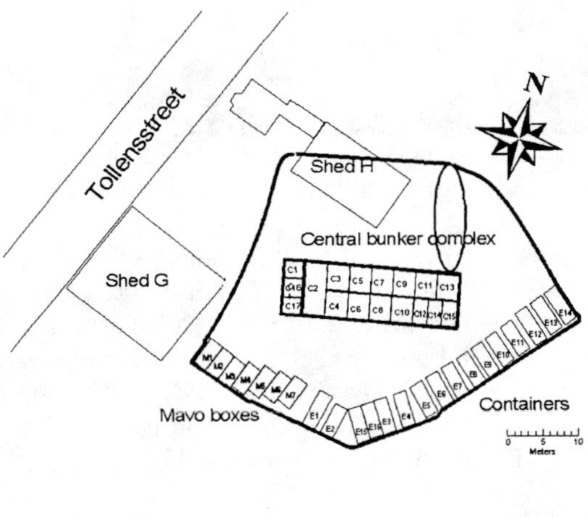

Figure 2: Flame from one of the storage units and the envelope indicating the individual risk contours of 10^{-5} per year and 10^{-6} per year.

SITUATION 2 – THE ACTUAL SITUATION

The probability of an accidental fire

In the actual situation, the probability of an accidental fire in a storage unit is the same as in the situation licensed, i.e. 8.8 $\times$ 10^{-4} per year per storage unit.

Consequences of the accident

In the actual situation, part of the fireworks has the properties of class 1.1 explosives and is therefore mass-explosive. In such a situation, it appeared that a fire in one of the storage units results in a number of explosions, finally destroying the complete storage facility. It is assumed that the explosions of the storage units are successive and the blast of the explosion of one storage unit does not interfere with the blast of the explosion of another storage unit. Therefore, the damage in the surroundings of the storage facility corresponds to the damage of the explosion of the largest storage unit. The mass detonation of the storage facility as a whole, leading to the simultaneous explosion of the whole content of the storage units, was not considered a realistic scenario, since at least part of the amount of fireworks stored was class 1.4.

The largest storage unit contained 7000 kg of fireworks, presumably class 1.1. The amount of explosive material in this type of fireworks is approximately 70%. Therefore, the effects of an explosion of a storage unit correspond to the explosion of approximately 5000 kg TNT. The sudden release of the explosion energy will give rise to high velocity fragments and a blast wave.

High velocity fragments.

Persons may suffer lethal effects if they are hit by a high velocity fragment generated in the explosion. The probability of suffering death from high velocity fragments depends on the density of the fragments and their impact on the body. Empirical relations are available to correlate the density of fragments at a certain distance with the strength of the explosion and the type of storage facility (Amelsfort, 1998). The density of fragments, q, as a function of the distance, R, is for the explosion of a not-igloo type of storage of 5000 kg TNT in a free field situation given in Appendix 1.

To determine the impact on the body, a distinction is made in three classes: fragments with mass between 0.001 kg and 0.1 kg having an initial velocity of 1000 m s^{-1} (50%) or 300 m s^{-1} (50%), fragments with mass between 0.1 kg and 4.5 kg having an initial velocity of 200 m s^{-1} and fragments with mass > 4.5 kg having an initial velocity of 100 m s^{-1}. Using the relations of Amelsfort (Amelsfort, 1998), at each distance R the distribution of the fragments over the classes is calculated and their velocity. Finally, probit functions are used to derive the probability of suffering death following a hit by a high velocity fragment due to skin perforation (criterion for mass class 0.001 kg – 0.1 kg), kinetic energy (mass class 0.1 kg – 4.5 kg) or fractured skull (mass class > 4.5 kg) (Amelsfort, 1998, TNO

1990). The probit relations are given in Appendix 1. The result of the calculation is given in Fig. 3 showing the lethality as a function of distance to the storage facility due to fragments in a free field situation.

The distance corresponding with 1% lethality due to fragments generated in the blast is equal to approximately 400 m. However, It should be noted, that the calculation is valid for a free field situation, i.e. a situation where no objects are present between the storage facility and the persons possibly hit. In the situation of the storage facility of S.E. Fireworks, the surroundings consist of a built up area and the persons are to a large extent protected against fragments. Therefore, in this situation fragments were not considered as contributing significantly to the probability of death. It appeared that one person died at a distance of 580 m of the storage by being hit by a fragment.

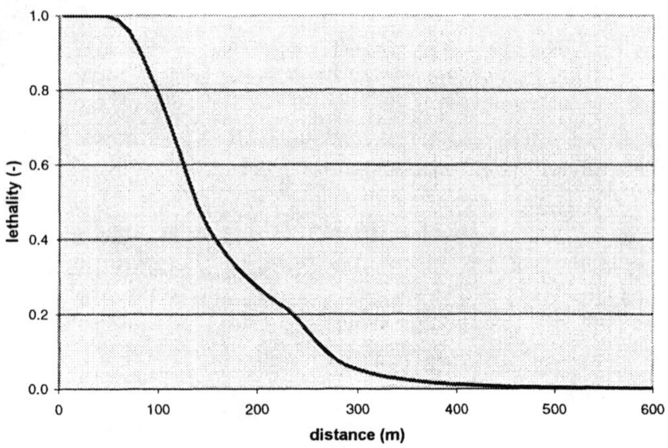

Figure 3: The lethality as a function of distance to the storage facility due to fragments in a free field situation, given an explosion.

Blast

Persons may suffer lethal effects by exposure to a blast wave, either directly by the overpressure or indirectly by the collapse of buildings. For the explosion of 5000 kg of TNT, the overpressure and impulse as a function of distance for light constructions are given by Amelsfort (Amelsfort, 1998) and are shown in Appendix 1. Applying these correlation functions and the probit functions for suffering death (see Appendix 1) shows that lethality due to the overpressure directly may occur at distances up to 40 metres.

However, the largest risk to persons in a built-up area is caused by the collapse of buildings. It is assumed that all persons (inside and outside) suffer death by an overpressure exceeding 0.3 bar due to the collapse of buildings and the resulting fragments. Between 0.1 and 0.3 bar all people outside survive, whereas 2.5% of the people inside suffer death due to the collapse of buildings (CPR, 1999). If the overpressure is less than 0.1 bar, no lethality occurs. Applying the correlation function for overpressure (see Appendix 1) shows that an overpressure of 0.3 bar can be found at a distance of 105 metres from the storage facility and an overpressure of 0.1 bar appears at a distance of 225 metres. The criterion used for suffering death due to the collapse of buildings, an overpressure of 0.3 bar, harmonises with the standard method of risk analysis (CPR, 1999).

Risk of the storage facility

The probability of an explosion is equal to 8.8×10^{-4} per year per storage unit. In the calculation of the individual risk, the probability of death within the 0.3 bar overpressure contour at 105 metres is equal to one and outside this contour is equal to zero. Hence the individual risk contours of 10^{-5} per year and 10^{-6} per year coincide and are given by the 0.3 bar overpressure contour as shown in Fig. 4.

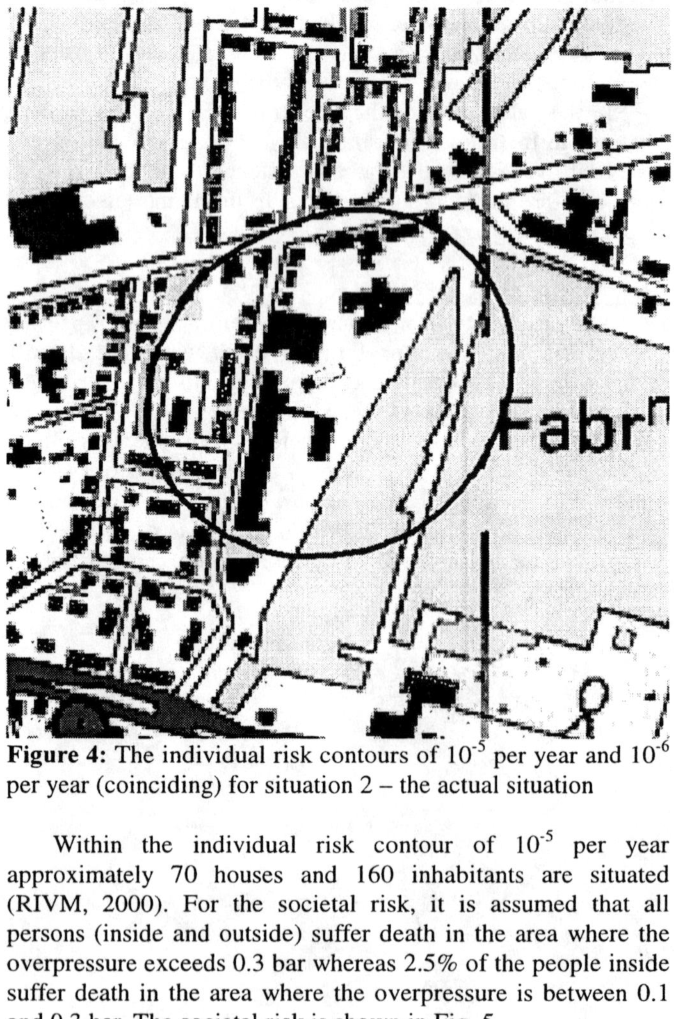

Figure 4: The individual risk contours of 10^{-5} per year and 10^{-6} per year (coinciding) for situation 2 – the actual situation

Within the individual risk contour of 10^{-5} per year approximately 70 houses and 160 inhabitants are situated (RIVM, 2000). For the societal risk, it is assumed that all persons (inside and outside) suffer death in the area where the overpressure exceeds 0.3 bar whereas 2.5% of the people inside suffer death in the area where the overpressure is between 0.1 and 0.3 bar. The societal risk is shown in Fig. 5.

DISCUSSION

The results of the calculation show that in case the establishment complied with the license, no risk is to be expected outside a 15 metres area around the storage units. The location of the facility would therefore comply with the criteria for individual risk and societal risk.

In the actual situation, the establishment does not comply with the individual risk and the societal risk criteria. Approximately 70 houses are situated within the individual risk contour of 10^{-5} per year and the indicative limit for the societal risk is exceeded by five orders of magnitude.

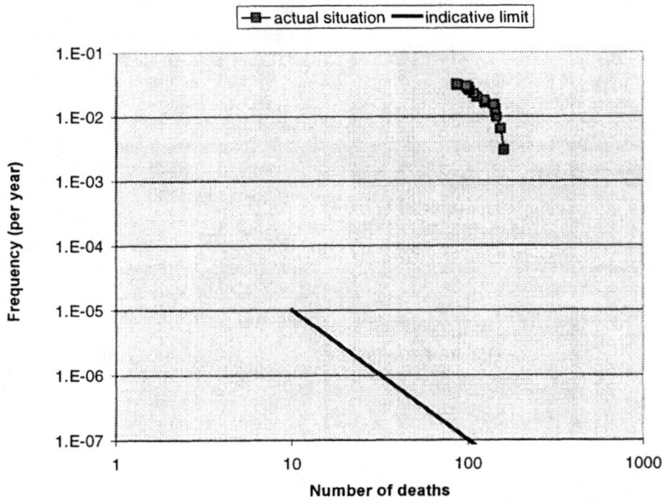

Figure 5: The societal risk for situation 2 – the actual situation, and the indicative limit

The results of the risk analysis of S.E. Fireworks can be compared with the risk calculations of the SEVESO establishments in the Netherlands (Laheij et al., 2000). In 1999, 124 establishments in the Netherlands were obliged to prepare a safety report. In total, about 70 persons are living within the individual risk contour of 10^{-5} per year of all the 124 establishments combined. Therefore, twice as many people were living within the individual risk contour of 10^{-5} per year of S.E. Fireworks than within the individual risk contour of 10^{-5} per year of all SEVESO establishments in the Netherlands.

Nine out of 124 SEVESO establishments exceed the indicative limit for the societal risk and the societal risk is maximally exceeded with a factor 52 (Laheij et al., 2000). The accumulated probability of an accident with 10 deaths or more is in the order of 10^{-4} in one year for all 124 SEVESO establishments together. The probability of an accident of S.E. Fireworks with 10 deaths or more is thus approximately a factor 100 larger than the probability of a similar accident in all 124 SEVESO establishments combined.

CONCLUSION

The results of the calculation show that in case the establishment complied with the license, the location of the facility would fulfil the criteria for individual risk and societal risk. In the actual situation however, the establishment does not meet the criteria for individual risk and societal risk. In fact, the establishment poses a larger threat to the society than all 124 SEVESO establishments in the Netherlands combined.

ACKOWLEDGMENT

The authors like to thank dr C. Twenhöfel for the calculation of the number of persons within the contours of the establishment.

REFERENCES

ADR, Accord européen relatif au transport international des marchandises dangereuses par route (European Agreement concerning the International Carriage of Dangerous Goods by Road).

Ale, B.J.M., 2001, "What caused the Enschede fireworks explosion", Safety Engineering and Risk Analysis 2001

Amelsfort R.J.M. van, 1998, "Description of the programs IRCONTOURS and FN_CURVES for the calculation of the individual risk and the societal risk in consequence of an accident at a storage of explosives", TNO, Rijswijk (in Dutch)

Bottelberghs, P.H., 2000, "Risk analysis and safety policy developments in the Netherlands", Journal of Hazardous Materials, 71, 59-84.

Commission to Investigate the Fireworks Disaster, 2001, "The Fireworks Disaster", Enschede/Den Haag

CPR. Committee for the Prevention of Disasters, 1999, "Guidelines for quantitative risk assessment", CPR18E, SDU, The Hague.

CPR. Committee for the Prevention of Disasters, 1990, "Methods for the calculation of damage (the Green Book)", CPR16E, Ministery of Social Affairs and Employement, The Hague.

Handbook of environmental licenses (updated continuously). ISBN 90-6500-215-4. Samson, Alphen aan den Rijn

Laheij, G.M.H., Ale, B.J.M., Post, J.G., 2000, "Population densities allowed around Dutch SEVESO establishments", SERA-Vol. 10, Safety Engineering and Risk Analysis 2000.

Minister of Housing, Spatial Planning and Environment, 1994, Letter of 13 June 1994 to the Second Chamber of Parliament, DGM/SVS/02594010 (in Dutch).

Minister of Housing, Spatial Planning and Environment, 1997, "Risk analysis method for CPR-15 establishments", Den Haag, (in Dutch)

RIVM, 2000, Database of houses and inhabitants in the Netherlands for 1999, Bilthoven

TNO, 2000, Experiments on a fire in a storage unit of fireworks on 14 October 2000.

APPENDIX I CORRELATION FUNCTIONS AND PROBIT FUNCTIONS

Correlation functions

In the risk analysis, correlation functions are used to determine the density of fragments, the blast overpressure and the impulse as a function of distance to the storage facility.

Density of fragments

The density of fragments q, as a function of the distance, R, is for the explosion of a not-igloo type of storage of 5000 kg TNT in a free field situation given by (Amelsfort, 1998):

$$\log q = 11.7633 - 5.08955 \times \log R \qquad \text{for } q \leq 0.5\ m^{-2}$$
$$q = 2.7493 \times 10^4 \times R^{-2} \qquad \text{for } q > 0.5\ m^{-2}$$

with q the density of fragments (m^{-2}) and R the distance to the storage facility (m).

Blast - overpressure

The overpressure, P_s, as a function of the distance, R, is for the explosion of a light construction storage of 5000 kg TNT in a free field situation for distances between 2 - 680 m given by the following relations (Amelsfort, 1998):

$$U = -0.214362789151 + 1.35034249993 \times LOG(0.05848 \times R)$$

$$LOG(P_s) = 2.78076916577 - 1.6958988741 \times U - 0.154159376846 \times U^2 + 0.514060730593 \times U^3 + 0.0988534365274 \times U^4 - 0.293912623038 \times U^5 - 0.0268112345019 \times U^6 + 0.109097496421 \times U^7 + 0.00162846756311 \times U^8 - 0.0214631030242 \times U^9 + 0.0001456723382 \times U^{10} + 0.00167847752266 \times U^{11}$$

with P_s the overpressure (kPa) and R the distance to the storage facility (m).

Blast - impulse

The impuls, I_s, as a function of the distance, R, is for the explosion of a light construction storage of 5000 kg TNT in a free field situation for distances between 20 - 680 m given by the following relations (Amelsfort, 1998):

$$U = -1.94708846747 + 2.40697745406 \times LOG(0.05848 \times R)$$

$$LOG(I_s) = -0.09419354 - 0.384519026965 \times U - 0.0260816706301 \times U^2 + 0.00595798753822 \times U^3 + 0.014544526107 \times U^4 - 0.00663289334734 \times U^5 - 0.00284189327204 \times U^6 + 0.013644816227 \times U^7$$

with I_s the impulse (kPas) and R the distance to the storage facility (m).

Probit functions

In the risk analysis, probit functions are used to calculate the probability of lethality due to exposure to fragments and the blast wave.

Density of fragments

The probit functions for the calculation of lethality due to fragments are for the different mass classes given by (Amelsfort, 1998):

$$Pr = -38.83 + 2.08 \times \ln (m \times v^{5.115}) \qquad \text{for } 0.001 \text{ kg} - 0.1 \text{ kg}$$
$$Pr = -17.86 + 5.36 \times \ln (0.5 \times m \times v^2) \qquad \text{for } 0.1 \text{ kg} - 4.5 \text{ kg}$$
$$Pr = -13.19 + 10.54 \times \ln (v) \qquad \text{for} > 4.5 \text{ kg}$$

with Pr the probit value, m the representative mass of a mass class (kg) and v the representative velocity of a mass class ($m\ s^{-1}$).

Blast

The probit functions for the calculation of lethality due to the overpressure and impulse are derived for lung damage, impact of the head and impact of the body (Amelsfort, 1998, CPR, 1990). For lung damage, the probit function is given by:

$$Pr = 5.0 - 5.74 \times \ln (4.2 \times (P_0/P) + 1.3 \times (P_0^{0.5} \times M^{1/3} / I)$$

with Pr the probit value, P the pressure (Pa), P_0 the ambient pressure (Pa), I the impulse (Pas) and M the mass of an adult (kg). For impact of the head, the probit function is given by:

$$Pr = 5.59 - 7.9 \times \ln (2.43 / P_s + 400 / (P_s \times I_s))$$

with Pr the probit value, P the overpressure (kPa) and I the impulse (kPas). For impact of the body, the probit function is given by:

$$Pr = 5.00 - 2.44 \times \ln (7.28 / P_s + 1300 / (P_s \times I_s))$$

with Pr the probit value, P the overpressure (kPa) and I the impulse (kPas).

Proceedings of
2001 ASME International Mechanical Engineering Congress and Exposition
November 11–16, 2001, New York, NY
SERA-Vol. 11

IMECE2001/SERA-24016

OUTLINE OF THE E&TM GROUP SYMPOSIUM "PARTNERING TO SUCCEED: KEYS TO MANAGING TECHNOLOGY DEVELOPMENT, RISK AND GLOBALIZATION"

Hiroshi Honda, Editor,
Proceedings of the Engineering and Technology Management Group Symposium
E-Mail Address: Hondah9876@aol.com (Honda International, Ltd.)
Telephone and Fax Number: 81 + 47 477 8571

The current paper introduces an outline of the subject symposium with **"twenty-six sessions,"** shown in Table 1, for ASME 2001 IMECE. (1)

DESCRIPTION OF SYMPOSIUM KEY CONCEPT LECTURES, FIVE TRACKS AND SELECTED SESSIONS

(Session A11/E&TM-9B: Symposium Key Concept Lectures: "Partnering to Succeed: Keys to Managing Technology Development, Risk and Globalization")

This session will discuss the symposium theme "Partnering to Succeed: Keys to Managing Technology Development, Risk and Globalization" and set scene for the entire discussion to be carried out during the symposium. Hiroshi Honda and Bjong "Wolf" Yeigh will cover various aspects of the theme from perspectives of industry, government and academia.

Track 1: Knowledge Learning, Protection and Transfer ***(Track Leader:*** *Susanne Tinker, Henderson, Finnegan, LLP (susanne.tinker@finnegan.com)* ***Track Sponsors:*** *Technology & Society Division (http://www.asme.org/divisions/ts/), Management Division (http://www.asme.org/divisions/management/)*)

Technology is revolutionizing how business is conducted. Businesses are becoming successful based on the intellectual capital they possess. Entrepreneurs are facing higher than ever success rates as businesses become regarded for their intellectual capital over physical structures. Intellectual capital in the form of intellectual property and human knowledge are the keys to achieving market position and to providing leverage with one's partners. These sessions will focus on developing technology partnerships and businesses, managing and protecting technology intellectual property, forming university/industry teams and consortia and developing technology transfer strategies. The following shows description for three sessions in this track:

(T12/E&TM-5A: Current Trends in Protecting Intellectual Property)

Technology businesses in today's world must learn how to develop, manage, and protect intellectual property in order to thrive. This session will address various industry perspectives and strategies for protecting intellectual property. Topics of discussion will include how to determine what type of protection is best suited for a particular kind of intellectual property, due diligence issues, insuring intellectual property, and other cutting edge issues pervading the ever-growing, ever-changing world of intellectual property.

(T-11/E&TM-6A: Technological Innovation Partnerships: Intellectual Property Rights Between Industry and Government)

This session will address the treatment of intellectual property rights for technology developed during, before and after certain types of partnerships between governmental agencies and industrial and/or academic entities. The primary focus will be jointly developed technology between companies and government representatives and the ownership and related treatment of intellectual property rights that arise under these relationships, including traditional government procurements, cooperative agreements, joint ventures, CRADAs and "other transactions."

Table 1 List of Twenty-Six Sessions Organized by Engineering and Technology Management Group

Track No.	*Track Title (Representative, Sponsoring Divisions)□ (Note) T&S: Technology and Society (Division)*			
Session No	Day	□	Time	Session Title (Organizer)
E&TM-9A □	Nov 12 □	Mon □	3:45 - 5:15pm □	Symposium Key Concept Lectures: "Partnering to Succeed: Keys to Managing Technology Development, Risk and Globalization" (Hiroshi Honda)
Track 1	***Knowledge Learning, Protection and Transfer (Susanne Tinker, T&S and Management Divisions)***			
E&TM-5A	Nov 12	Mon	7:45 - 9:15 am	Current Trends in Protecting Intellectual Property
E&TM-6A □	Nov 12 □	Mon □	9:30 - 11:00 am □	Technological Innovation Partnerships: Intellectual Property Rights between Industry and Government
E&TM-7A	Nov 12	Mon	11:15 - 12:45 pm	Managing Technology Transfer and Partnering
E&TM-8A	Nov 12	Mon	2:00 - 3:30 pm	Life-Long Learning as the Engineer's Responsibility to Self
Track 2	***Ethics, Business and Engineering (Steve Long, T&S and Management Divisions)***			
E&TM-7B	Nov 12	Mon	11:15 - 12:45 pm	Did I do the Right Thing? (Ethics, Business, and Being a Good Engineer)
E&TM-8B	Nov 12	Mon	2:00 - 3:30 pm	Ethics World: Corporate, Professional, Individual
E&TM-9B	Nov 12	Mon	3:45 - 5:15 pm	Risk Communication
Track 3	***Safety & Risk, Rising to the Challenge (Ben Ale, Safety Engineering and Risk Analysis Division)***			
E&TM-13	Nov 13	Tue	2:00 - 3:30 pm	The New Frontier of Product Safety at the End of Product Life
E&TM-14A	Nov 13	Tue	3:45 - 5:15 pm	The Global Application of Risk-informed or Performance-based Standards
E&TM-15	Nov 14	Wed	7:45 – 9:15 am	Impacts of Standards and Design on Safety
E&TM-16	Nov 14	Wed	9:30 - 11:00 am	Beneficial Approaches and Examples of Hardware Maintainability for Safer Use
E&TM-17A	Nov 14	Wed	11:15 - 12:45 pm	Business Approaches to Managing Risk and Reliability
E&TM-18	Nov 14	Wed	3:45 – 5:15 pm	Conflict Resolution between Siting of Facilities and the Public
Track 4	***Sustainable Design & Evolving Manufacturing Methods (Theodore Aanstoos, Management Division)***			
E&TM-1 □	Nov 11 □	Sun □	11:15 - 12:45 pm □	Product Sustainability through Design, Management, and Public Policy -Can We Afford It?
E&TM-2	Nov 11	Sun	2:00 - 3:30 pm	Factory 2005: Strategies, Practices, and Tools for Agile Manufacturing
E&TM-3	Nov 11	Sun	3:45 - 5:15 pm	Failure Analysis & Prevention
E&TM-4	Nov 11	Sun	5:30 - 7:00 pm	Entrepreneurship: The Engineer as a Profit Center
E&TM-5B	Nov 12	Mon	7:45 - 9:15 am	New Engineering Business: Planning for Growth
Track 5 □	***Collaboration, History and Advancement of Technology and Education, and Their Impact on Society and Industries (Hiroshi Honda, Technology and Society Division)***			
E&TM-6B □	Nov 12 □	Mon □	9:30 - 11:00 am □	Past President's Perspective: Modern History of Technology and Its impact and Interaction with Society and Industries (1)
E&TM-10A □	Nov 13 □	Tue □	7:45 - 9:15 am □	Advancement of Materials Technology and Its Implication and Impact on Future Society and Industries
E&TM-10B	Nov 13	Tue	7:45 - 9:15 am	Collaborative Engineering
E&TM-11 □	Nov 13 □	Tue □	9:30 - 11:00 am □	Advancement of Space Utilization and Its Implication and Impact on Future Society and Industries
E&TM-12 □	Nov 13 □	Tue □	11:15 - 12:45 pm □	Past President's Perspective: Modern History of Technology and Its Impact and Interaction with Society and Industries (2)
E&TM-14B □	Nov 13 □	Tue □	3:45 - 5:15 pm □	Advancement of Energy and Engineering Technology and Its Implication and Impact on Future Society and Industries
E&TM-17B □	Nov 14 □	Wed □	2:00 - 3:30 pm □	Modern History of Mechanical Engineering Education and Its Interaction with Society and Industries

(T-13/E&TM-7A: Managing Technology Transfer and Partnering)

This session will address various industry perspectives and strategies for achieving successful technology transfer arrangements and relationships. Topics of discussion will include how to select and find a suitable partner, how to finance the partnership, and how to protect both parties' rights to the developed and transferred technology. Current and expected trends and cutting edge methods relating to technology transfer will also be presented.

Track 2: Ethics, Business and Engineering

*(**Track Leader:** Steve Long (Steve.Long@gdseng.com) **Track Sponsor:** Technology & Society Division (http://www.asme.org/divisions/ts/), Management Division (http://www.asme.org/divisions/ management/))*

Globalization of markets and evolving business strategies are changing the rules of the game. Traditional issues such as licensing, regulations, supply chain, business policy and ethics are being revised and companies are scrambling to catch up. This session will look at how globalization is changing the expectations and demands on industry and what business must do to stay ahead.

Track 3: Safety & Risk, Rising to the Challenge

*(**Track Leader:** Ben Ale (ben.ale@rivm.nl), National Institute for Health and Environment, The Netherlands **Track Sponsor:** Safety Engineering & Risk Analysis Division (http://www.serad.org))*

Safety, risk, and reliability are key public issues, as seen in public reports about transportation, accidental releases, and manufacturing. The Safety Engineering and Risk Analysis Division (SERAD) is focusing on new and evolving codes, standards, and assessment and management methods as applied to industry. Sessions in this track will focus on areas such as: risk-based standards and regulations, product safety (after life, retirement), maintainability (RAM), success via reliable and maintainable systems, performance-based standards, transport safety and siting issues.

(S-31/E&TM-13: The New Frontier of Product Safety at the End of Product Life)

There is increasing demand for environmentally friendly and safe disposal of products after their economic life. This means that the concern for the safety of products not only exists during a product's lifetime but goes beyond into the period of taking products back with storage of disposed products and their dismantling. For new products, this leads to additional demands put on the safe design and construction of a product, which gives rise to new challenges early in a product's life cycle and long before its end of life. For older or existing products, the challenge occurs at the end of life. This session will discuss the motivation, issues, problems, and solutions of product safety after the end of life of a product.

(S-52/E&TM-14A: The Global Application of Risk-Informed or Performance–Based Standards)

The development of standards with an increasingly larger, multinational coverage is the consequence of the globalisation of markets. These standards therefore are more and more used outside the immediate area of influence of the organisation(s) where the standards were developed or implemented. This means that the risk of misinterpretation, the lack of data, or deficiencies in the implementation of the standards have to be identified and managed with increasing care. This session will provide an update to risk-related global standards and the problems and their solutions to their uniform application within a global company or across national boundaries.

(S-34/E&TM-15: Discussion Session: Impacts of Standards and Design on Safety)

On the one hand the number of products and services increases, on the other hand society seems less and less tolerant to risk. In this light a structured approach to safety issues across technologies seems to be needed. Standards and quality managed design procedures have a place in this approach. The issues related to the role of standards and design in the promotion of safety are the subject of a panel discussion.

(S-32/E&TM-16: Beneficial Approaches and Examples of Procedures for Safer Use)

The increasing demand for the reduction of waste also leads to the desire to have manufacturing equipment, process vessels, or products last longer. This means that inspection, maintenance, and repairs have to be performed on hardware or products and parts of hardware and products that previously were disposed of much earlier. This not only holds for consumer products such as household appliances but also for airplanes, chemical plant equipment, and even nuclear power plants. When products are already in use, the design cannot be altered and construction changes that are part of maintenance present new challenges to ascertain continued safe use. This_session will discuss current and evolving approaches to maintainability for equipment or products and their safe use.

(S-33/E&TM-17A: Business Approaches to Managing Risk and Reliability)
Reliability and maintainability not only contribute to safety but also have considerable economic advantages. These advantages include longer production runs for a process, greater life span for a product and lower disposal costs. In most cases, the benefits outweigh the disadvantages incurred due to the decrease of sales of new products, greater "upfront" cost of more reliable equipment, and the continued use of older technology. This session will discuss the metrics for success for aging as well as new hardware or systems, how systems are monitored for reliable performance, and the approaches that can be taken for a successful outcome using reliable and maintainable systems.

(S-35/E&TM-18: Conflict Resolution between Siting of Facilities and the Public)
The volume of production of dangerous goods and the associated transport or shipment of these goods increases in the industrial world to accommodate expanding and changing economies. The mode of transport also changes from traditional means to transport by container, pipeline and new underground structures. On the other hand, the concentrations of population increase in both size and intensity leading to increasing conflicts between the demand for space for the production and transport facilities and the demand for safe distances between these facilities and the population. Several ways of solving these conflicts are tried in the world and this session will discuss the metrics, approaches, and issues to solving these complex societal risk problems.

Track 4: Sustainable Design & Evolving Manufacturing Methods ***(Track Leader:*** *Theodore Aanstoos (t.aanstoos@mail.utexas.edu), The University of Texas at Austin* ***Track Sponsor:*** *Management Division (http://www.asme.org/divisions/management/))*

Sustainable development issues have risen to the forefront of management concerns, especially in the face of increasing globalization and changing regulation. Life-cycle management is being addressed by many companies, which have never before dealt with such issues. This session will discuss the changes this focus on sustainable development is causing, including impacts on regulation and government, DFX, management strategies, and third world development. The manufacturing community is rapidly adapting to dramatic changes in the management of supply chains and new technologies for life cycle designs. Focus on systems, outsourcing and e-commerce are replacing traditional structures. This session will focus on agile/lean manufacturing, e-delivery, life cycle design and outsourcing, including novel outsourcing methods such as virtual factories.

Track 5: Collaboration, History and Advancement of Technology and Education, and Their Impact on Society and Industries ***(Track Leader:*** *Hiroshi Honda, Honda International, Ltd. (Hondah9876@aol.com)* ***Track Sponsor:*** *Technology & Society Division (http://www.asme.org/divisions/ts/))*

Globalization of markets and business strategy are driving rapid change in how companies (and nations) do business. Politics, global and sustainable development, and increasing global competition and collaboration are all driving business and government into new partnerships.

This track will examine international trends in collaborative engineering, history and advancement of technology and education, and their impact on society and industries.

Particular focus will be placed on the energy, materials, space, communications, information and other critical technologies and industries, and mechanical engineering education. Table 2 shows the entire presentations for Track 5 and key concept lectures sessions, and the following shows description for sessions in this track:

(T-56 and -57/E&TM-6B and E&TM-12: Past President's Perspective: Modern History of Technology and impact and Interaction with Society and Industries (1) and (2))
ASME Past Presidents Leroy S. Fletcher (1985-86), Daniel T. Koenig (1995-96), and Winfred M. Phillips (1998-99) will give a presentation on "The History of Heat Transfer," "The Role of the Engineer in the 21st Century," and "Bioengineering from Mechanics and Devices to Tissue Engineering and Genetics," respectively at Session (1).

Nathan H. Hurt (1991-92) and Keith B. Thayer (1997-98) will give a presentation on "Nuclear Technology: A National Treasure" and "History of Technology 1950-2000: Engineering Changes Observed," respectively at Session (2). All their presentations are based on their life-long achievement and experience in their own fields of pursuit.

(M-51/E&TM-10A: Advancement of Materials Technology and Its Implication and Impact on Future Society and Industries)

Vijay K. Varadan, Wole Soboyejo, and Oludele Popoola will give presentations on microsensors, MEMS and NEMS based systems, and automotive materials.

(T-55/E&TM-10B: Collaborative Engineering)

Carolyn Schlueter, et al., and Dong Hongzhao, et al., will give presentations on "Virtual Teams: Communicating in a Global Business" and "Small-Medium-Enterprises (SMEs) Oriented Industrial Alliance," respectively.

Discussions include issues such as miscommunications in international collaboration the enterprise behavior protocol and systematic architecture of Industrial Alliance.

(T-54/E&TM-11: Advancement of Space Utilization and Its Implication and Impact on Future Society and Industries)

Lawrence J. DeLucas, NASA Astronaut (June 1992 Space Shuttle Launch), will discuss on "Protein Crystal Growth in Space, Past and Future" and Po K. Wong will give a presentation on "Basic Needs of Human Beings as the Purposes and Foundations for the Existence of Governing Institutions and the Advancement of Science and Technology."

(T-53/E&TM-14B: Advancement of Energy and Engineering Technology and Its Implication and Impact on Future Society and Industries)

Arthur L. Baldwin, Srinivas Garimella, Jeffrey J. Krimmel, et al., and John K. Gershenson, et al., will give presentations on fossil energy technology for tomorrow with a focus on the role of effective government partnerships, challenges and opportunities for space conditioning systems, carbon tax issues in the age of a move toward renewable energies, and production decision making issues in the face of uncertainty about the air toxics regulation and global climate change policy, respectively.

(T-15/E&TM-17B: Modern History of Mechanical Engineering Education and Interaction with Society and Industries)

Shigehiko Kaneko, et al., John M. Howell, and Eva Loven, et al., will give presentations on project based learning in the area of micro-gas turbine, use of project based education and interactive Web resources in undergraduate heat transfer, and modeling and simulation used to understand learning of new computer-based tools, respectively.

AKNOWLEDGEMENT

The author would like to thank Kenneth Horne, Susanne Tinker, Theodore Aanstoos, and Ben Ale for providing him with the session materials.

REFERENCES

(1) Honda, H., Partnering to Succeed: Keys to Managing Technology Development, Risk and Globalization and Appendix A, Proceedings of Symposium on Partnering to Succeed: Keys to Managing Technology Development, Risk and Globalization, (ed. H. Honda), IMECE 2001, ASME (November 2001)

Table 2 List of Sessions, Chairs, Presentations and Presenters for Symposium Key Concept Lectures and Track 5 Sessions

Session/E&TM No. Chairs	**Day** **Day**	□ **Week**	**Time Zone** Presenters, Affiliation	**Session Title** Presentation Title
T-56/E&TM-6B	**Nov. 12**	**Mon**	**9:30 - 11:00am**	**Past President's Perspective: Modern History of Technology and Its impact and Interaction with Society and Industries (1)**
Arthur E. Bergles			Leroy S. Fletcher NASA Ames Research Center	The History of Heat Transfer □
Hiroshi Honda			Daniel T. Koenig Consultant	The Role of the Engineer in the 21st Century □
			Win M. Phillips Univ. of Florida	Bioengineering from Mechanics and Devices to Tissue Engineering and Genetics
A-11/E&TM-9A	**Nov. 12**	**Mon**	**3:45 - 5:15pm**	**Symposium Key Concept Lectures: "Partnering to Succeed: Keys to Managing Technology Development, Risk and Globalization"**
Hiroshi Honda			Hiroshi Honda Honda International, Ltd.	Partnering to Succeed: Keys to Managing Technology Development, Risk and Globalization (1)
Steven P. Nichols			Bjong "Wolf" Yeigh Yale Univ.	Partnering to Succeed: Keys to Managing Technology Development, Risk and Globalization (2)
M-51/E&TM-10A	**Nov. 13**	**Tue**	**7:45 - 9:15am**	**Advancement of Materials Technology and Its Implication and Impact on Future Society and Industries**
Hiroshi Honda			Vijay K. Varadan Penn State Univ.	Microsensors, MEMS and NEMS based Complex Adaptive Smart Devices and Systems
			Wole Soboyejo Princeton Univ.	Recent Advances in MEMS and Bioengineering Technology: Implications for Future Society
			Oludele Popoola Ford Research Laboratory	Recent Advances in Automotive Materials □
T-55/E&TM-10B	**Nov. 13**	**Tue**	**7:45 - 9:15am**	**Collaborative Engineering**
Shuichi Fukuda			Carolyn Schlueter Michigan State Univ.	Virtual Teams: Communicating in a Global Business □
			Dong Hongzhao Zhejiang Univ., P. R. China	SMEs Oriented Industrial Alliance □
T-54/E&TM-11	**Nov. 13**	**Tue**	**9:30 - 11:00am**	**Advancement of Space Utilization and Its Implication and Impact on Future Society and Industries**
Hiroshi Honda			Lawrence J. DeLucas Univ. of Alabama	Protein Crystal Growth in Space, Past and Future □
			Po K. Wong Systems Research Company	Basic Needs of Human Beings as the Purposes and Foundations for the Existence of Governing Institutions and the Advancement of Science and Technology

Table 2 (Continued)

Session/E&TM No.	Day□		Time Zone	Session Title
Chairs	Day	Week	Presenters, Affiliation	Presentation Title
T-57/E&TM-12	**Nov. 13**	**Tue**	**11:15 - 12:45pm**	**Past President's Perspective: Modern History of Technology and Its impact and Interaction with Society and Industries (2)**
Paul J. Torpey			Nathan H. Hurt Technical and Management Consulting	□Nuclear Technology: A National Treasure □
Richard E. Rosenbeg			Keith B. Thayer Garuda U. S., Inc.	History of Technology 1950-2000 Engineering Changes Observed□
T-53/E&TM-14B	**Nov. 13**	**Tue**	**3:45 - 5:15pm**	**Advancement of Energy and Engineering Technology and Its Implication and Impact on Future Society and Industries**
Hiroshi Honda			Arthur L. Baldwin National Energy Technology Lab.	Fossil Energy Technologies for Tomorrow: The Role of Effective Government Partnerships for Sustainable Progress
Theodore A. Aanstoos			Srinivas Garimella Iowa State Univ.	Novel Energy Efficient and Environmentally Friendly Space-Conditioning Systems: Challenges and Opportunities
			Jeffrey J. Krimmel Univ. of Texas	Paradigm Shift: A Move toward Renewable Energies □
			John K. Gershenson Michigan Technological Univ.	Production Decision Making in the Face of Uncertainty about Air Toxics Regulation and Global Climate Change Policy
T-15/E&TM-17B	**Nov. 14**	**Wed**	**2:00 - 3:30pm**	**Modern History of Mechanical Engineering Education and Its Interaction with Society and Industries**
Hiroshi Honda			Shigehiko Kaneko Univ. of Tokyo	Project Based Learning Taking Micro Gas Turbine as a Theme
			John M. Howell Univ. of Texas	Using Project-Based Education and Interactive Web Resources in Undergraduate Heat Transfer
			Eva Lovén Linköping University	Modelling and Simulation Used to Understand Learning of New Computer- Based Tools -in cooperative and competitive engineering context-

AUTHOR INDEX

SERA-Vol. 11
Safety Engineering and Risk Analysis — 2001

Book Number: I00524